SEP 2002

363.1246 ADAIR
Adair, Bill.
The mystery of Flight 427.

P9-DWE-924

10|8

WITHDRAWN

HVA ... TUM

THE MYSTERY OF
FLIGHT427

FLIGHT427

THE MYSTERY OF
FLIGHT427
INSIDE A CRASH INVESTIGATION

BILL
ADAIR

SMITHSONIAN INSTITUTION PRESS
WASHINGTON LONDON

ALAMEDA FREE LIBRARY
2200-A CENTRAL AVENUE
ALAMEDA, CA 94501

© 2002 Bill Adair
All rights reserved

Copy Editor: Jan McInroy
Production Editor: Ruth G. Thomson
Designer: Jody Billert

Library of Congress Cataloging-in-Publication Data
Adair, Bill.
    The mystery of Flight 427 : inside a crash investigation / Bill Adair.
        p.   cm.
    Includes bibliographical references and index.
    ISBN 1-58834-005-8 (alk. paper)
        1. USAir Flight 427 Crash, 1994. 2. Aircraft accidents—Investigation—
    United States. 3. Aircraft accidents—Pennsylvania—Hopewell
    (Beaver County : Township) I. Title.
    TL553.525.P4 A33   2002
    363.12.465.0974892—dc21   2001049601

British Library Cataloging-in-Publication available

Manufactured in the United States of America
09 08 07 06 05 04 03 02   5 4 3 2 1

⊚ The paper used in this publication meets the minimum requirements of the
American National Standard for Information Sciences—Permanence of Paper
for Printed Library Materials ANSI Z39.48-1984.

For permission to reproduce illustrations appearing in this book, please correspond
directly with the owners of the works, as listed in the individual captions. The
Smithsonian Institution Press does not retain reproduction rights for these
illustrations individually or maintain a file of addresses for photo sources.

ALAMEDA FREE LIBRARY
2200-A CENTRAL AVENUE
ALAMEDA, CA 94501

For Katherine

CONTENTS

The important thing to understand about the rudder pedals is that they are unnecessary; like your wisdom teeth, they serve no very good purpose but can cause much trouble.

Wolfgang Langewiesche, 1944

PROLOGUE

A BAD DREAM

Summer 1995

Great Falls, Virginia

The clock on the nightstand read 2 A.M., and Tom Haueter was wide awake. He was usually a leaden sleeper, dead to the world once his head hit the pillow. But tonight a nightmare had jolted him awake.

By day Haueter ran the investigation into the crash of USAir Flight 427. He was the consummate man in charge, all confidence and certainty. At night, though, his doubts sometimes overcame him. It had been nine months since the Boeing 737 corkscrewed out of the blue sky over Pittsburgh and dived into a hill at 300 miles per hour, but Haueter still didn't know why it had happened.

He had run many investigations for the National Transportation Safety Board (NTSB), and this one had started like all the rest—the peculiar smell of death mixed with jet fuel and the adrenaline rush during the first few days of examining the wreckage. But the rush he received had long since passed. Investigators usually figure out the cause a week or two after a crash, but not this time. They had eliminated one theory after another—the promising ones, the far-fetched ones, and a few that were truly bizarre—and now it seemed they were back where they had started.

At the NTSB, solving a case was paramount. It was right there in federal law: "The board shall report the facts, conditions and circumstances relating to each accident and the probable cause thereof." If an investigator couldn't come up with the cause, he had failed. In the entire twenty-five-year history of the NTSB, only four cases had gone unsolved—and one of those involved a 737.

Indeed, Haueter was accustomed to solving every case, even seemingly impossible ones like the crash in Brunswick, Georgia, that killed U.S. senator John Tower. That case was especially difficult because the evidence was so sketchy. The Embraer 120 plane did not have a flight data recorder or a cockpit voice recorder. Haueter had to rely on radar data and the pilots' last words with air traffic controllers. But ultimately the NTSB had found a piece of wreckage, no bigger than a coin, that revealed a flaw in the propeller system.

Many of Haueter's colleagues at the board believed that he would never find the answer to the USAir crash. Some thought he should give up. "You've got nothing," one investigator said. "It's time to walk."

Maybe it was. He fantasized about quitting. He was fed up with the office politics and the childish sniping between Boeing and the pilots union. He certainly could live without his beeper, without the calls in the middle of the night, and without his job stealing his weekends. It would be good for his marriage.

But his fantasies about quitting didn't last long. He realized there was no way he could leave in the middle of the biggest mystery in NTSB history. His personal and professional pride was on the line. And there were lives at stake.

Haueter's bosses were putting immense pressure on him. One of them said that if the USAir case went unsolved, Congress would abolish the NTSB. If the bozos at the board couldn't solve this one, Congress would say, they might as well find a new line of work. That was just the kind of pressure Haueter didn't need. He not only had to figure out whether the world's most widely used jetliner had a fatal flaw, he also had to save his agency from extinction.

When he complained to a friend about the pressure, NTSB chairman Jim Hall got wind of it and intercepted Haueter one day as he was walking out of the office.

"Can you follow me down for a minute?" Hall asked.

They walked through the lobby, past the NTSB's Most Wanted List of safety recommendations, and entered an elevator.

"You know, you don't have to solve it," Hall said.

"Jim, I appreciate that, but yeah, we do," Haueter replied. "Greg and I need to know what happened. We don't want this thing hanging over us. We've got four unsolved accidents. The safety board can't afford a fifth."

Haueter felt that he *could* solve it. He just needed time. But he worried about what might happen in the meantime.

In his nightmare, another 737 had crashed, which prompted Congress to launch a massive inquiry to find out why the NTSB had bungled the case.

Suddenly he was in a giant hearing room, facing a panel of angry congressmen. The TV cameras were zooming in on him. It seemed there were thousands of people in the room, and all of them had decided he was guilty. He was at the witness table, all alone.

"What happened?" a congressman demanded. "Why didn't you do something sooner? Why didn't you ground the fleet?"

1

A GOOD AIRPLANE

September 8, 1994

Lisle, Illinois

Brett and Joan Van Bortel pulled into the parking lot shortly after sunrise, with a few minutes to spare before Joan's 6:20 train. She was a marketing manager for Akzo Nobel, a big chemical company, and liked to get to work early, even when it meant a twelve- or fourteen-hour day. This would be one of those days. She was flying to Pittsburgh for a dinner meeting.

Joan took a trip nearly every week and had become a seasoned business traveler. She carried the same suitcase-on-wheels that pilots and flight attendants used, and she traveled light, taking only the bare essentials for each trip. Unfortunately, the prime spots for the chemical business were not the nation's most glamorous cities. She spent a lot of time in Akron, Ohio, the rubber capital of the world.

Joan was an ambitious person. Her goal was to become Akzo Nobel's highest-ranking woman. She was one of the first people to arrive in the office each morning and usually ate lunch at her desk so she could keep working. She told her employees that every call should be picked up by the third ring. She was not a chemist, but she took time to learn about the company's products. She held training sessions to teach employees how to pronounce the chemical names and made them take written quizzes with questions like "How is rubber cured?" and "Name one of our products that has zinc in it." When Joan stopped at a gas station, she got into long conversations with auto mechanics about the chemistry of tires.

As she and Brett kissed good-bye, Joan looked very professional in a stylish green-and-white suit, with her briefcase in hand. She wore her engagement ring, which had a distinctive marquise diamond surrounded by other diamonds. Brett had given her lots of jewelry, but this ring was her favorite.

She was five feet two—almost a foot shorter than Brett—with shoulder-length honey-brown hair, sparkling brown eyes, and a flawless smile. With her hair up, she resembled the actress Jessica Lange. Brett loved the way Joan was comfortable with a grunge look—big glasses, a baseball cap, and messy hair—and the way she could transform herself into a knockout. She exercised every day and was in great shape, which allowed her to indulge in an occasional bag of Skittles from the office snack machine.

While Joan was in Pittsburgh that night, Brett planned to stay home and install a tile floor in their kitchen. He had promised her the floor would be finished before she came home the following day.

Captain Peter Germano and First Officer Charles B. Emmett III first saw Ship 513 in Jacksonville, Florida. They had spent the previous two days flying to Indianapolis, Philadelphia, Toronto, Cleveland, Charlotte, and then down to Jacksonville, a trip that involved three different 737s. Switching planes during a trip was standard procedure for most airline pilots. Because of union work rules and government time limits, USAir pilots flew no more than eight hours per day, followed by a mandatory nine hours and fifteen minutes of rest. A typical USAir 737 was in the air for ten hours every day, however. That timing mismatch led to a complex and confusing schedule, as the airline tried to maximize productivity by switching crews on and off different aircraft throughout the day.

Ship 513 had spent the night in Windsor Locks, Connecticut, where mechanics performed a "transit check" on the plane, inspecting the hydraulic system for leaks, examining the wheels and tires, and checking the engine oil. There were no significant maintenance problems or pilot "squawks" that needed to be fixed. On the morning of September 8, a different set of pilots had flown the plane from Windsor Locks to Syracuse, Rochester, Charlotte, and then Jacksonville. Emmett and Germano were scheduled to take the aircraft to Charlotte, Chicago, and Pittsburgh.

Ship 513 was identical to the 220 other 737s that USAir flew. It had a shiny silver fuselage (painted planes were heavier, which meant higher fuel costs) and a pair of stripes, red and blue, running the length of the plane just below the windows. The tail was navy blue with red pinstripes and the airline's simple logo in white letters. The company colors were also featured on the inside—the seats were navy blue with red and white decorations. The bulkhead that separated coach from first class was covered in a carpet that looked like a sunset. It was supposed to absorb sound so that people talking in coach wouldn't bother the first-class passengers.

The first-class section had 8 leather seats, and coach had 118 fabric-covered ones, each designed to be as thin and lightweight as possible and still comply with federal safety standards to withstand a forward force of nine times the force of gravity, or 9 Gs. It was a sharp-looking plane, a big improvement over USAir's previous colors, frumpy 1970s earth tones that one USAir official had described as "red on brown on red on brown."

Ship 513 was seven years old, which made it a relatively new plane in the USAir fleet. Purchased in October 1987 for about $24 million, the plane had logged 23,800 hours—the equivalent of flying continuously for nearly three years. It had made almost 14,500 flights or "cycles"—the most critical measurement of a plane's age. Each time an aircraft is pressurized for a flight, the airframe is subjected to stress.

The plane was part of the 300 series, which meant it was the third generation of 737s. The first generation, the 100 series, was introduced in 1967. Boeing designed the 737-300 to be in service for at least 75,000 cycles, but many planes continued to fly long after that. The life span was economic, ending when it became too costly to maintain the planes. Airlines typically kept jets for twenty to thirty years before trading them in for new models.

Emmett, Germano, and the flight attendants had arrived in Jacksonville about 11 P.M. on September 7 and checked in to the Omni Jacksonville Hotel, a downtown high-rise overlooking the St. John's River. Germano ordered a turkey croissant sandwich from room service shortly before midnight and called his wife, Christine, back in Moorestown, New Jersey. He and his crew would be able to sleep late the next morning; they didn't have to be back at the Jacksonville airport for their next trip until noon.

Their flight to Charlotte was uneventful.

On the next leg, to Chicago, a USAir pilot named Bill Jackson rode in the cockpit jump seat, a fold-down seat behind the pilots. It was common practice in the airline industry to allow pilots to ride for free so they could commute from their home city to their crew base. Many pilots preferred riding in the jump seat so they did not have to listen to annoying chatter from passengers.

About thirty minutes into the flight to Chicago, Andrew McKenna Sr., a passenger in first class, heard a strange gurgling sound. He was a seasoned traveler, the head of a major paper and packaging company, so he was accustomed to the noises inside big jets. But this was unusual, like water being forced out of a sink. It seemed to be coming from just above his head.

McKenna summoned the flight attendant and described the noise. He flew a lot, he told her, and he had never heard anything like it. She listened for a moment and then said she thought the sound was coming from the PA speaker. McKenna wasn't sure that she was right, but he went back to his reading. He didn't give the sound another thought.

The flight attendant picked up the intercom phone, called Germano in the cockpit, and reported that a passenger was hearing an unusual noise that seemed to be coming from the PA system. Germano turned to Jackson behind him in the jump seat and noticed that his knee was pressing on a microphone button. Jackson moved his knee, and the flight continued to Chicago without further complaints.

But after the plane landed at O'Hare and parked at Gate F6, passengers were still talking about the noise. The gate was packed with people waiting to board the plane for its next trip, Flight 427 to Pittsburgh. As the arriving passengers walked off, a woman whose husband was booked on Flight 427 overheard someone discussing the noise. She decided to call USAir to make sure her husband's plane was safe.

The phone rang in the mechanics lounge beneath the F gates just as USAir maintenance foreman Gerald Fox walked in the door. He listened as the woman explained what she had overheard. She said she was concerned because her husband was on the Pittsburgh flight.

"I have two good mechanics on duty," Fox told her. "If there is a problem with the airplane, it will be taken care of before leaving." He hung up and walked outside to look for his mechanics. After searching for a few minutes, he walked up the metal stairs to Ship 513, which was being loaded for the trip to Pittsburgh. He found Germano in the covered Jetway just outside the plane and explained the woman's complaint about the strange noise. Germano did not mention the microphone incident from the previous flight, but he did not seem concerned.

Germano said, "I have a good airplane."

At the Akzo Nobel office in downtown Chicago, Joan had gotten busy with meetings and phone calls, so she was running late when she grabbed her bags at 3:45 P.M. and ran out of the office to catch the El train to O'Hare. She had only an hour and fifteen minutes to get to the airport, and several people at the office thought she would miss her flight. She had decided to take along a laptop computer so she could work on a report at her Pittsburgh hotel. This was the first time she'd carried a laptop, and she was worried that it might get zapped by the airport metal detector, but her coworkers assured her it would be fine.

Joan got off the train at O'Hare and dashed through the underground tunnels and up the escalator to Terminal 2. She tossed her suitcase and briefcase on the X-ray belt, walked through the metal detector, and grabbed her bags. She hurried past the shoeshine stand and the snack bar to Gate F6. She would have preferred to fly American or United—where she had most of her frequent flier miles—but neither of those airlines had many flights to Pittsburgh. Her travel agent had booked her on USAir, which had a big hub there.

At O'Hare, however, USAir was a bit player. The airline's gates in the F wing looked like they hadn't been improved since the days of the first Mayor Daley. Under yellowed ceiling tiles, passengers sat in cramped gray chairs and watched the CNN Airport Network on a blaring TV. The hallway echoed with the sound of footsteps and the *clickety-clickety-clickety* of suitcase wheels. The PA system kept telling passengers: "May I have your attention, please. For security reasons, keep your baggage with you at all times. Unattended baggage will be removed by the Chicago Police Department." A red cardboard sign told passengers to watch for suspicious activity and to refuse packages from "anyone you do not know very well."

Joan handed her ticket to the agent and walked down the Jetway toward Ship 513. She was in 14E, a middle seat just behind the wing. Though she preferred to sit on the aisle, nothing was available there. Flight 427 was packed. In the seat on her right was Robert Connolly, a financial consultant headed home to Pittsburgh. In the one on her left was a man from Virginia named John T. Dickens. The plane was so full that the Weavers, a family of five from Upper St. Clair, Pennsylvania, had to sit in middle seats scattered around the cabin. Seven-year-old Scott Weaver was one row ahead of Joan, and his eleven-year-old sister, Lindsay, was one row back. The family was returning from a funeral for a nine-year-old cousin.

It was primarily a business flight. Eight U.S. Department of Energy employees were returning to Pittsburgh from a coal conference. Several of them had initially booked seats on later USAir flights but had switched to this one so they could get home earlier. Also on board were four people from US Steel, a lawyer from Westinghouse, and an account executive from a Chicago radio station. The man in 20C was a neuroscientist from the Scripps Institute for Oceanography. The grad student in 16A was flying to Pittsburgh for a job interview. The well-tanned guy with the baseball cap in 17F was a convicted drug dealer.

At the gate, Captain Germano was given the flight plan, the weather forecast, and the cargo manifest on a computer printout that stretched four feet long. Pilots often joked about the big stack of paperwork for each flight, saying that when the weight of the paper exceeded the weight of the airplane, it was safe to fly. The papers told Germano that Flight 427 was scheduled to leave at 4:50 P.M. Chicago time and land in Pittsburgh 55 minutes later. The plane would have a cruising altitude of 33,000 feet and would get a gentle push from a 31-knot tailwind. The plane would need 6,400 pounds of jet fuel, but it would carry more than twice that amount in case Germano had to divert to another city or go into a holding pattern.

The plane's route looked like gibberish: ORD . . GIJ.J146 . . J34.DJB . . ACO . CUTTA1 . PIT, but Germano could read it like simple street directions. The three-letter codes stood for airports and navigation markers between Chicago and Pittsburgh. Flight 427 would climb away from O'Hare (ORD), over a point known as Gipper east of Gary, Indiana, and then up to jet routes J146 and J34. They were like interstate freeways in the sky, carrying high-altitude east-west traffic along the Indiana-Michigan border and then southeast toward Pittsburgh. Flight 427 would cross over a navigation point known as Dryer near Cleveland and then begin to descend near Akron, Ohio. It would follow a standard arrival route known as "CUTTA," which was like a big funnel for planes from the northwest converging on the Pittsburgh airport.

Several pages of Germano's paperwork dealt with the weather. There were SIGMETS—significant meteorological conditions—for Georgia and Florida, but none that would affect his brief flight over the Midwest. The weather in Pittsburgh looked perfect, sunny skies with temperatures in the mid-seventies. All of the Pittsburgh runways were dry.

The papers also gave Germano crucial information about the plane's weight and the speed necessary to get off the ground. The plane would be carrying 11 tons of people and 3,700 pounds of cargo. The aircraft and its contents would weigh 115,000 pounds as it roared down the O'Hare runway. It would need to go at least 138 knots, or 159 miles per hour, to get airborne.

At the bottom of the main page, Germano saw this statement:

I HEREBY ACKNOWLEDGE RECEIPT OF THIS FLIGHT PLAN AND NECESSARY ATTACHMENTS AND CONSIDER ALL CONDITIONS INCLUDING MY PHYSICAL CONDITION SUITABLE FOR THIS FLIGHT. I HAVE ADEQUATE KNOWLEDGE OF ALL FACTORS AFFECTING THE ROUTE, WEATHER, NAVIGATION, COMMUNICATIONS, TERRAIN, OBSTRUCTIONS AND ALL APPLICABLE PROCEDURES AND REGULATIONS.

Germano printed his name and his USAir employee number, then signed his name. The lives of 131 people on board were now his responsibility.

As the passengers stuffed their carry-on luggage into the overhead bins, baggage handlers filled the belly of the plane with 1,700 pounds of luggage and a ton of *BusinessWeek* magazines that were ultimately headed to subscribers in the Carolinas. The flight was running about fifteen minutes late, so USAir mechanic Tim Molloy had extra time to walk around the plane and make sure it was safe. He circled Ship 513 twice, checking the tires, the wings, the rudder, the tubes that measure airspeed, and the fluid levels for the hydraulic systems. He made sure all the cargo doors were locked. No problems. The plane looked fine.

Either Molloy or mechanic Mark Kohut pushed the plane back with a USAir tractor—neither of them remembers who performed which task—and told the pilots by intercom that it was safe to start the engines. The mechanic then stood away from the plane and snapped the pilots a salute. Flight 427 was on its way.

2

Z U L U

The cockpit in Ship 513 was identical to every other USAir 737. With pilots switching planes two or three times a day, it was crucial that the instruments and controls be in exactly the same place in all the planes. The cockpit seemed to be filled with a hundred clocks. It was possible to equip 737s with more modern computer screens, but USAir chose to stick with the older-style "steam gauges" so that all its planes would be standardized.

The walls and panels of the cockpit were gray, a neutral color that allowed pilots to see the dials more easily. There were two seats with sheepskin covers, for the captain and the first officer, plus the fold-down jump seat behind them that could be used by Federal Aviation Administration inspectors or pilots hitching a ride from one city to another. A sign on the back wall of the cockpit said: LIQUOR TAX HAS BEEN PAID, a requirement because the airline served alcohol. The cockpit door had a small mirror on the inside so the pilots could straighten their hats and ties before saying hello or good-bye to passengers.

Many people who earn more than $100,000 a year have spacious offices, but not airline pilots. They work in a room smaller than a bedroom closet. The 737 cockpit is a familiar, comfortable place to them, however. The controls and instruments are laid out very logically. The most important controls are directly in front of the pilots—the rudder pedals and the wheel/control column. The most important gauges—airspeed, altitude, the attitude indicator, and the compass/navigational dial—form a T in the center of each pilot's instrument panel. Switches and levers that are used less frequently are placed farther away. The circuit breaker panel, which is not used very often, is directly behind the seats. Above the pilots' heads is a small compartment with an escape rope so they can climb out a window and slide down the fuselage if the cockpit door is blocked. Hidden beneath the jump seat is an ax, which pilots can use to chop into an electrical panel during a fire or to break out of wreckage after a crash.

By standardizing cockpits, the airlines are encouraging repetition. If pilots perform the same task repeatedly, it should become so automatic that they don't make mistakes. That's also the rationale behind requiring them to use checklists—to make sure that they flip each lever the same way, in the same order, on every flight. Checklists are no guarantee that a crew won't screw up—the checklists themselves can become so rote that pilots race through them without doing what the list calls for—but when used properly they provide a good tool for helping the pilots go through the tasks consistently. Standardization is crucial because captains and first officers may never have flown together before. Pilots pick their trips based on their own personal schedules and their favorite cities, so the selection of a copilot is usually just a matter of luck. (A prized USAir trip was Baltimore–St. Thomas–St. Croix–Baltimore, which had a 25-hour overnight at a nice resort in St. Croix; the least popular were the red-eyes, such as the 2 A.M. Los Angeles–to–Pittsburgh flight.)

Germano and Emmett had been through the 737 checklists thousands of times and could probably have recited them from memory. But before the plane departed, the pilots were still required to go through them point by point. The lists had a unique rhythm, like a rap song with two singers alternating back and forth:

> Fuel quantity?
> 15-6 required; 15-6 on board.
> Oil and hydraulic quantities?
> Checked and checked.
> Fuel panel?
> Set.
> Seat belt sign?
> On.
> Window heat?
> On.
> Hydraulics?
> A's off; B's on.
> Pressurization?
> Set.

On each flight the captain and first officer trade off the tasks of flying the plane and communicating with air traffic controllers, thus spreading the workload evenly and assuring that they both get a chance to fly. But a distinct pecking order is still in effect. The captain, whose uniform carries four stripes on the shoulder epaulets, has the ultimate responsibility. If the captain thinks that anything about the plane is unsafe, the flight won't leave. Likewise, only the captain has the authority to abort a takeoff.

The top job comes with a few perks. USAir 737 captains made about $160,000 a year in 1994, whereas first officers made $110,000. Also, the captain traditionally gets to sit inside the cockpit while the first officer performs the walk-around inspection outside the plane, which can be a miserable task during rain or snow.

Germano, from Moorestown, New Jersey, was forty-five years old and had been flying since he was seventeen. He flew for the New York State Air National Guard and began his airline career with Braniff Airways in 1976. He started with USAir in 1981, initially as a flight engineer on the Boeing 727, then as first officer on the BAC-111 and then as first officer and captain on the 737. He was an accomplished pianist, had been married for nineteen years, and had two daughters, ages three and nine.

Emmett, who was thirty-eight, also began flying as a teenager. He started his career by flying corporate planes and in 1987 joined Piedmont Airlines, which was bought by USAir two years later. He was married and lived in the Houston suburb of Nassau Bay. He loved to sail, and he drove a Corvette with the Texas license plate 1USAIR. At six feet four, he was one of the tallest USAir pilots.

It was Emmett's turn to fly, so Germano would be handling radio duties on the leg to Pittsburgh. Assuming that they followed standard airline procedures—as virtually every USAir pilot did—the taxiing and takeoff would have gone like this:

After one of the mechanics pushed the plane back with a tractor, Emmett turned the ignition switch and Germano moved a lever to start the No. 2 engine, the one on the right wing. After waiting about forty seconds for the engine to spool up, the pilots started No. 1. Germano moved a lever to engage the parking brake until they were cleared to leave. Emmett set the flaps on the wings to provide the extra lift it would take to get the plane airborne. (It is crucial to set the flaps. Two crashes in the late 1980s were the result of pilots' forgetting to set them.)

The pilots then went through the "After Start" checklist, making sure that the generators and hydraulic pumps were on and the engine anti-ice was set properly. They checked for heat in the pitot tube, a sensor that measures airspeed, and then checked their shoulder harnesses to make sure they were snapped and secure.

"After-start checklist complete," said Emmett.

Ground controllers cleared them to follow the taxiways until they reached Runway 32-Left, where they waited for another controller's direction. It was shortly before 5 P.M. Central Time.

"Cleared for takeoff," the controller said.

Emmett moved the throttle levers forward and pushed a button marked "TOGA," which stood for "take off/go around." That action energized the autothrottles so the plane's flight-management computer would control the big CFM-56 engines. The computer would keep the power steady as the plane climbed.

The silver 737 began to roll down the runway. Just as Emmett removed his left hand from the throttle levers, Germano placed his right hand on them. It would be Germano's responsibility to decide whether to reject the takeoff.

"Eighty knots," Germano called out.

Emmett looked at his airspeed indicator to make sure it agreed. "Checked," he responded.

The plane was nearing V-1, the speed at which it could no longer be stopped on the runway.

"V-1," said Germano, removing his hand from the power levers. They were committed now. They had to fly.

"Rotate," said Germano.

Emmett pulled back on the control column, lifting the plane's nose into the Chicago sky. They were airborne.

"Gear up," Emmett said.

Germano grabbed the gear lever—it had a small wheel on the end so it would be unmistakable—and flipped it up. The pilots heard a thump as the nose gear was pulled inside the plane.

Emmett relied on the autopilot most of the way. USAir wanted its crews to use the device as much as possible because it made the plane more fuel-efficient. It was like cruise control in a car. Emmett could set the desired airspeed, altitude, and heading on a panel just below the windscreen, and the plane would automatically follow that course. Ship 513 also had a flight-management computer that kept track of the plane's route and position and told the autopilot when to turn, climb, or descend. The computer could be cranky, however. About thirty minutes into the flight, Emmett had trouble getting it to accept a command.

"Ah, you piece of shit!" said Emmett.

"What?" asked Germano.

"I said, 'Aw, c'mon, you piece of shit!' This damn thing is so fucking slow!"

Emmett cursed the computer twice more before it did what he wanted. "There it is," he said, finally satisfied.

The plane was at 29,000 feet as it cruised along the Michigan-Indiana border and then over the sparkling waters of Lake Erie, before banking gently to the right and turning southeast toward Cleveland.

"USAir 427, cleared direct to Akron, rest of route unchanged," a controller in Cleveland told Germano. "Give me the best forward airspeed, in-trail spacing."

"Direct Akron, best forward, you got it," Germano said. "USAir 427."

They began a steady descent toward 24,000 feet. Once they reached that point, another controller told them to continue down to 10,000 feet, the point where they would enter the CUTTA arrival pattern into the Pittsburgh airport.

Germano tuned the radio to the recorded weather briefing. "Pittsburgh tower arrival information Yankee," it said. "Two-one-five-two Zulu weather. Two five thousand scattered. Visibility one five. Temperature seven five. Dew point five one. Wind two seven zero at one zero."

That meant the weather was ideal: 75 degrees Fahrenheit with scattered clouds and 15 miles of visibility. It was a perfect summer evening. The two pilots were relaxed. It was a Thursday shortly before 7 P.M. Eastern time, their last day of work that week. They chatted with a flight attendant about pretzels and sampled her fruit juice–Diet Sprite concoction.

"That's good," said Germano after taking a sip.

"That is different," said Emmett. "Be real, be real good with some dark rum in it."

"Yeah, right!" the flight attendant said.

The plane had crossed Ohio and was nearly to the Pennsylvania state line as it steadily descended toward 10,000 feet. About this time, the flight attendants were probably walking through the cabin to collect cups and cans. Passengers were told to put away computers and other electronic gadgets that might affect the plane's navigational equipment.

"USAir 427, Pittsburgh Approach," air traffic controller Richard Fuga told the pilots. "Heading one-six-zero, vector I-L-S Runway two-eight Right final approach course. Speed two-one-zero." Fuga sounded as if he was in a great mood. His voice was playful as he directed planes toward the airport.

The pilots had been told to slow the airspeed to 210 knots and fly a heading toward Runway 28-Right. The plane was closing in on the Pittsburgh airport now, and Germano had listened to the latest radio briefing on airport conditions, which was known as "Yankee."

"We're coming back to two-one-zero," Germano replied to Fuga. "One-sixty heading down to ten, USAir 427 and, uh, we have Yankee."

A minute later, Fuga told them to descend to 6,000 feet. Germano acknowledged it, saying, "Cleared to six, USAir 427." The pilots went through a preliminary checklist, making sure that the altimeters and other flight instruments were set properly.

"Shoulder harness?" Germano asked.

"On," replied Emmett.

"Approach brief?"

"Plan two-eight-right, two-seven-nine inbound, one-eleven-seven." They had set the navigation radios to align the plane with the runway.

Ship 513 was the last plane from the northwest in a big wave of arrivals. After landing in Pittsburgh, it would continue to West Palm Beach. But Emmett and Germano would switch to yet another 737 and fly the final leg of their trip across Pennsylvania to their home base, Philadelphia.

Fuga told Germano to slow the plane to 190 knots and begin turning toward the Pittsburgh airport at a compass heading of 140.

Germano acknowledged, saying, "Okay, one-four-zero heading and one-nine-zero on the speed, USAir 427."

One of the pilots switched on the seat belt sign, but then Emmett realized he hadn't told the passengers to prepare for landing. "Oops, I didn't kiss 'em 'bye. What was the temperature, 'member?"

"Seventy-five."

"Folks, from the flight deck, we should be on the ground in 'bout ten more minutes," Emmett announced over the PA system. "Uh, sunny skies, little hazy. Temperatures, temperatures ah, seventy-five degrees. Wind's out of the west around ten miles per hour. Certainly appreciate you choosing USAir for your travel needs this evening, hope you've enjoyed the flight. Hope you come back and travel with us again. At this time we'd like to ask our flight attendants, please prepare the cabin for arrival. We'd ask you to check the security of your seat belts. Thank you."

Germano was confused about the runway assignment. "Did you say two-eight Left for USAir 427?" he asked the controller.

"Uh, USAir 427, it'll be two-eight Right," Fuga said.

"Two-eight Right, thank you."

Germano then listened to Fuga slow other planes to 190 knots, the equivalent of 218 miles per hour. "Boy, they always slow you up so bad here," he said to Emmett.

"That sun is gonna be just like it was takin' off in Cleveland yesterday, too." Emmett said, laughing. "I'm just gonna close my eyes. You holler when it looks like we're close."

Germano chuckled. "Okay."

They were about four miles behind Delta Air Lines Flight 1183, a Boeing 727 that was going to land ahead of them. Another plane, an Atlantic Coast Airlines Jetstream commuter plane, had just taken off and was about to enter their area.

"USAir 427, turn left heading one-zero-zero. Traffic will be one to two o'clock, six miles, northbound Jetstream climbing out of thirty-three for five thousand," Fuga told them. The commuter plane was headed from 3,300 feet to 5,000, but it would stay miles away.

"We're looking for the traffic, turning to one-zero-zero, USAir 427," said Germano.

They started a gentle left turn. "Oh, yeah," Emmett said, mocking a slight French accent, "I see zuh Jetstream."

"Sheeez," said Germano.

"Zuh," said Emmett.

Thump. The plane suddenly rolled to the left. *Thump.*

"Whoa," said Germano. The wings on the big 737 started to level off, but now the left wing rolled down again.

"Hang on, hang on," Germano said. Emmett grunted.

One of them clicked off the autopilot, triggering the *whoop-whoop-whoop* of the autopilot warning horn.

"Hang on," said Germano.

"Ohhh shiiiiit," Emmett said in his slight Texas twang, sounding increasingly worried.

To passengers back in the cabin, the bumps initially felt like routine tur-bulence. But then the plane kept rolling left, and the nose pitched down to-ward the ground.

The pilots were desperately trying to figure out what was happening. One of them pulled back on the control column, trying to get the nose up.

"What the hell is this!!?" Germano exclaimed. Moments earlier, he had been able to see the horizon and a perfect blue sky. Now all he could see was the ground. Only twelve seconds had passed since the first hint of trouble.

The cockpit was chaotic. Stickshakers on the pilots' control columns began rattling like jackhammers, warning them that the plane was stalling. The autopilot warning kept blaring *whoop-whoop-whoop*, notifying them that it had been disconnected. But that was the least of their problems. The plane's traffic computer spotted the Jetstream a few miles away and its electronic voice shouted "TRAFFIC! TRAFFIC!"

"What the . . . !!!?" asked Germano.

The plane was still a mile up in the sky above the Green Garden Plaza shopping center, diving straight down at 240 miles per hour, twisting like a leaf and gaining speed. Out their front window, the pilots could see trees, roads, and the shopping center spinning closer and closer. As the plane cork-screwed down, passengers were pressed back in their seats by centrifugal force so strong that they had difficulty even lifting their hands off their laps. The wings had been robbed of their ability to fly, which made the plane shake vio-lently, as if it were running over a thousand potholes.

"Oh!" said Emmett.

"Oh God! Oh God!" cried Germano.

The dials and gauges in the cockpit spun like clocks rushing forward in time. Germano shouted to controllers, "Four-twenty-seven emergency!"

The plane continued to dive toward a rocky hill.

"Shit!"

"Pull!"

They were only 700 feet above the hill and diving at 280 miles per hour.

"Oh shit!"

"Pull!"

"God!" cried Emmett.

Germano screamed, "Pullllllllll!"

It had been just twenty-eight seconds since the first inkling of trouble.

Just before impact, Emmett sounded resigned, almost pleading, as he said, "Noooo . . ."

In the eerie darkness of the Pittsburgh TRACON, a windowless room filled with glowing radar screens, Richard Fuga saw the plane's altitude suddenly drop to 5,300 feet.

"USAir 427, maintain six thousand," he told them. "Over."

He heard "emergency" and the pilots' final cries. Either Emmett or Ger-mano had kept his finger on the radio button as the plane fell.

The altitude on Fuga's radarscope suddenly changed to three *X*s. That meant the plane was falling so fast that the FAA computer did not believe it. A moment later the plane disappeared from the screen.

Fuga called to them urgently. "USAir 427, Pittsburgh?"

No response.

"USAir 427, Pittsburgh?"

Still nothing.

Fuga gave rapid directions to another pilot and then called again for the missing plane. "USAir 427, Pittsburgh?"

Nothing.

He then said sadly, "USAir 427 radar contact lost."

He asked other controllers to take over his flights and summoned a supervisor. He pointed to his screen. "Last radar and radio on 427, right here."

Dozens of people saw the USAir plane fall. It was 7:03 P.M. in Hopewell Township, and the soccer games were in full swing on a field a few blocks from the hill. The 737 had flown over the soccer field and then rolled left and plunged toward earth.

"Look at that airplane!" shouted someone on the field.

In a car a mile south of the soccer field, Mike Price saw the plane twist out of the sky. "That airplane's in trouble," he told his father. It looked like someone had picked up the 737 by its tail and let it fall straight down. In the parking lot at Green Garden Plaza, Amy Giza had just climbed into her car and was reading the directions for a new set of math flash cards when her six-year-old son said, "Mommy, that airplane just fell out of the sky."

George David, the owner of a 62-acre farm on Green Garden Road, was cutting flowers in his yard when he heard the roar of the plane's engines. He thought it might be a truck racing out of control. Then he heard the explosion as the plane struck the gravel road that led to his neighbor's house. Trees blocked everyone's view of the actual impact, but lots of people saw the fireball erupt a moment later. Inside the Giant Eagle grocery store at Green Garden Plaza, the crash sounded like a huge crack of thunder.

A plume of smoke rose from the hill and drifted across Route 60, over the Beaver Lakes Golf Course. At least seventy-five people called 911. The first person to reach the Hopewell Township police department—entered into the log as "hysterical caller"—said a plane had crashed behind the shopping center. At fire stations throughout the Pittsburgh area, firefighters heard a series of tones and then "Zulu at Pittsburgh International Airport." A Zulu call meant a disaster with at least twenty people killed.

More than forty fire trucks, ambulances, and police cars raced to the crash site, about ten miles west of the airport. When Engine 921 of the Hopewell Volunteer Fire Department reached the woods at the top of the hill, Captain James Rock hopped out and grabbed an ax and a pry bar. He was a professional firefighter at a nearby Air Force base and had taken part in many drills rescuing people from plane crashes. He dashed through the woods, ready to pry passengers out of the wreckage and save some lives.

He saw mangled luggage and airplane seats. He saw a man's dismembered hand on the ground. He looked around feverishly. There was no one to save.

Firefighters pulled hoses into the woods and sprayed water on the wreckage and the trees to douse the flames. Others ran through the woods, shouting for survivors.

"Anybody here?!" they yelled. "Anybody need any help?!"

There was no reply.

A police officer stood at the center of the debris, right where the nose had hit, and asked, "Where's the plane?"

Down the hill at the shopping center the scene quickly became chaotic. Dozens of fire trucks and ambulances showed up, even though Hopewell Township authorities had not requested them. When fire chiefs and ambulance drivers throughout the Pittsburgh area heard there had been a plane crash, they just piled into their trucks and drove to Hopewell, eager to help.

They were not needed. There were a few fires to put out, and there was plenty of need for police to direct traffic and protect the crash site, but rescuers in dozens of ambulances and advanced life support trucks had nothing to do. This would be a cleanup operation, not a rescue call.

At FAA headquarters in Washington, a phone rang in the operations center on the tenth floor. It was the FAA nerve center for crashes, terrorism, and other mayhem, a place that looked like a remnant from the Cold War. In one room was a sophisticated TV-computer system that allowed the ops officer to watch all four major TV networks simultaneously. In another corner was a big radio panel with microphones and dials that looked like something out of *Dr. Strangelove*. It let the FAA communicate with airports and air traffic controllers if telephones got knocked out in a hurricane or a military attack.

The conference room next door served as a situation room, a place where FAA officials could plot strategy in a crisis and be in constant touch with people around the country. The phone system allowed elaborate conference calls for up to 240 people. That was especially useful after a crash, when the FAA wanted to link its accident investigators with their counterparts from the National Transportation Safety Board so they could make arrangements to travel to the site.

Ops officer Sharon Battle took the call about Flight 427 from someone in the FAA's northeast regional office. She then pulled out the gray "Notification Record" that listed each office and government agency she needed to call. One by one, she went down the list, calling FAA administrator David Hinson and the rest of the FAA top brass, as well as the White House Situation Room, the FBI, and the CIA.

"We'd like to give you a briefing," she told each of them. "USAir Flight 427, a Boeing 737, O'Hare to Pittsburgh at 6,000 feet. Radio and radar contact lost. Unknown fatalities or survivors at this time. Unknown if any ground injuries."

She then made a round of calls to the accident investigators from the FAA and the NTSB. It was time to mobilize the Go Team.

John Cox and Bill Sorbie were in Pittsburgh for a USAir program called Operation Restore Confidence, a safety campaign about pilot mistakes and the need for pilots to follow procedures. The program had been in the works for months but had gotten new urgency because of the crash of USAir Flight 1016 in Charlotte two months earlier. It was USAir's fourth fatal crash in five years, which had prompted the FAA to scrutinize the airline to make sure it had no systemic safety problems. Wind shear had thrown Flight 1016 to the ground, but NTSB investigators were likely to blame the pilots for flying into the storm.

Cox and Sorbie were USAir pilots and safety officials with their union, the Air Line Pilots Association (ALPA). The union had surprisingly good relations with the company, especially on safety issues. A recent joint program was a case in point. The airline was having repeated problems with pilots who strayed from their assigned altitude, which not only could be dangerous but also could lead to FAA fines. So ALPA and the company agreed on a new procedure in which both pilots were required to call out their assigned altitude and then point their index finger at the altitude number on the instrument panel. That simple routine had reduced the number of deviations by more than 90 percent. The union and the airline hoped that Operation Restore Confidence would have the same kind of dramatic effect. The six-hour program began with statistics about mistakes by USAir pilots and then discussed how they could improve and standardize their procedures.

Like many pilots, Cox and Sorbie had chosen to live in Florida, where taxes and housing prices were low, and commute to their crew base (Baltimore for Cox, Philadelphia for Sorbie) when they had to fly a trip. Cox and his wife, Jean, a USAir flight attendant, lived in a waterfront home in St. Petersburg. Sorbie lived on a houseboat a few miles away in Tierra Verde. The Operation Restore Confidence meeting ended in the late afternoon, but the pilots decided to stay and have dinner at Mario's, their favorite Italian restaurant, instead of rushing back to Florida. They finished dinner and were heading to their hotel with Don McClure, another ALPA official, when Sorbie's pager went off, followed by McClure's and then Cox's. Within a minute, the three pagers sounded again.

"Oh, shit," Sorbie said.

They got back to their hotel, a Hampton Inn at the airport, and went to Sorbie's room. He called the ALPA official who had paged them and got the news: A USAir plane was down.

"This cannot be," said Cox, who was still a member of the investigation looking into the crash of 1016. "This cannot be happening again." He figured it wasn't really a USAir jet. It was probably a USAir Express commuter plane. People often got the facts wrong in the first few hours after a crash.

But as the details emerged over the next hour it was clear that the plane was indeed a USAir 737, the same type of plane that Cox piloted.

They drove to the ALPA office near the Pittsburgh airport and spent ninety minutes on the phone notifying other people from the union and talking about which accident investigators should be summoned to the crash site.

They were about ten miles away themselves, so Cox, Sorbie, and another ALPA official arranged for a police escort and headed west on Route 60 toward Hopewell Township. They showed their USAir ID badges to police officers at several checkpoints and then drove up Green Garden Road and parked in a driveway. As they climbed out of their rental car, they saw smoke still coming from the hill. They borrowed flashlights from an officer, walked under a line of police tape, and picked their way through the trees. A firefighter came up to them and asked, "Who are you guys?"

"We're accident investigators from the pilots union," they said.

"There's not much here," the firefighter said.

The pilots asked if there were any survivors.

"No, nobody will get out of this one."

The first things Cox and Sorbie saw were some of the lightest items from the plane—EXIT signs and life jackets. As they got closer, they began seeing body parts and then larger pieces of the plane. Sorbie was struck by the lack of smells. After a plane crashes, there's usually the sweet aroma of jet fuel. Sorbie sniffed the air but couldn't smell it. *Geez,* he thought, *I hope the pilot didn't run the damn thing out of gas.*

It was a surreal scene. The plane appeared to have crashed on a long dirt road, but debris had been blasted in every direction. There were no lights on the road, so the fire department had brought in portable lamps that sprayed the trees with a harsh white light and cast long shadows in the woods. Fires were spontaneously popping up in the trees, and firefighters ran over to extinguish them.

As the pilots got closer to the road, they noticed larger and larger pieces of wreckage, but most were no bigger than a car door. It looked like the 109-foot-long plane had disintegrated. They walked all around the woods, shining their flashlights on the largest pieces of wreckage. The engines were battered but still whole. The biggest piece was the tail, but it was badly banged up. As they walked carefully around the road and through the woods, Cox kept looking for parts from a second airplane, figuring that the 737 had been in a midair collision. But as he aimed the flashlight at the hundreds of pieces on the ground and in the trees, he saw only fragments of the big silver jet.

"Seen enough?" asked Sorbie.

"Yeah," Cox said. "I've seen way more than enough."

3

NEXT-OF-KIN ROOM

Brett Van Bortel's company, Reed Elsevier, depended on business travelers like his wife, Joan. The company published the *Official Airline Guide*, known in frequent flier shorthand as the OAG, which listed complete airline schedules for every city in the country. Brett wrote brochures and magazine ads that portrayed the OAG as the bible of frequent travelers. He had even stirred up trouble with a billboard he had written. It stood just outside the entrance to O'Hare and read, O'HARE AHEAD. CARRY PROTECTION, with a picture of the *OAG Pocket Flight Guide*. City officials were not amused at the implication that people might need protection in their beloved airport, so the sign came down.

Brett was a child of suburbia. He and his two brothers grew up in West Chicago. His father was an executive with a food service company, his mom was a teacher. They lived in a spacious colonial house across the street from a picturesque forest preserve that had ponds and hiking trails. It was like an extension of the Van Bortels' front yard—a huge place where Brett and his brothers could build forts and go camping. In the winter they went cross-country skiing through the tranquil forest; on the Fourth of July, they climbed to the top of an old landfill called Mount Trashmore and watched the fireworks.

Brett was on the track and swimming teams and played middle linebacker and center on the freshman football squad. He broke his neck in a bad car accident when he was sixteen, but recovered completely. He had always been the writer in the family, even as a boy. On his eighteenth birthday, an age when many boys are in full rebellion against their parents, he wrote his mother a sentimental poem about how much he loved her. He chose the University of Iowa because it had a great English department. His favorite writers were classic authors—Thomas Hardy, Jonathan Swift, and Shakespeare. But he also liked *First Blood,* the book that was the basis for the Rambo movies.

Joan had grown up on a farm in Melrose, Iowa, a tiny town about sixty miles south of Des Moines. Melrose was known as "Iowa's Little Ireland" because most of its residents, including Joan's family, were Irish. Her parents grew corn and soybeans and raised cows. As the only girl in a family of five boys, Joan was spared most of the farm duties. That was just as well because she gradually discovered that she preferred living in the city. In choosing to go to the University of Iowa, Joan effectively said good-bye to farm life. (They say in Iowa that you go to the University of Iowa for culture and to Iowa State for agriculture. Joan had chosen culture.)

Joan and Brett were acquaintances for several years in college but did not start dating until their senior year. After they graduated, they spent a winter skiing in Vail, Colorado, and then moved to Chicago to start their careers. They had bought a ranch-style house on Riedy Road just before they were married. It was a fixer-upper with purple and green walls that desperately needed to be repainted. But they found it a lot more inviting than the sterile shoebox homes in nearby Naperville, the ones on streets with names like Whispering Woods, even though there wasn't a single native tree for miles. The Lisle house would take some work, but they could give it personality. They were not do-it-yourselfers, but figured they could learn. Their first project was the bathroom. They gave it a new coat of paint and wallpaper, and Brett replaced the toilet himself.

His latest project was installing floor tile in the kitchen. He had just placed the last tile when the phone rang. It was Joan's secretary.

"There's been a plane crash," she said. "I think Joan was on it."

Brett flipped on CNN. The first words out of the television were

" . . . no survivors."

"Oh, my goodness," said CNN anchor Linden Soles. "Well, we had initial reports of 123 people aboard, possibly 130 if that's counting a crew of 7. Are there a large number of emergency crews in the area right now, Sandra?"

"The whole county has responded—helicopters, ambulances, medi-rescue, police from all over the county," the woman replied.

"Now, your estimation that there are perhaps no survivors from this crash—is that based on what you've seen or have you heard any confirmation from any emergency personnel?"

"We have not really had any confirmation on it, but our understanding is that there are no survivors, but we are not confirmed on that."

Brett quickly dialed the number that CNN listed for USAir, but he kept getting busy signals. When he finally got through, the USAir employees were clueless. Brett said he thought his wife was on the plane that crashed. A USAir agent promised to have someone call back.

Brett's brother Grant had come over to help tile the floor. He could see that Brett was upset. "What's up?" he asked.

"I think Joan might have been in a plane crash," Brett said. His words came out matter-of-factly; it was foolish to jump to conclusions, right? He didn't know that she was on that particular plane. There were lots of flights from Chicago to Pittsburgh. What were the odds that she was on the plane that had crashed?

Joan's secretary said she would go to the office and check Joan's itinerary. In the meantime, Brett called Joan's credit card company, hoping that she had charged the tickets and they would have the flight number. The company was no help. Then he tried calling Bob Henninger, the coworker Joan was supposed to meet in Pittsburgh. He left Henninger a message and then repeatedly called the hotel where Joan was supposed to stay. But the hotel operator kept telling him she had not checked in yet. Brett called again and again. Finally the operator connected him to a room. The phone rang.

Thank God! thought Brett. *She's alive!*

The operator came back on the line: "I'm sorry. She hasn't checked in."

Minutes after the accident, a USAir supervisor typed a few commands into a computer to prevent anyone at the airline from seeing information about Flight 427. Reservation agents who tried to call up the passenger list got a curt response on their screen: UNABLE TO DISPLAY.

Copies of the passenger list were printed for only three locations— USAir's situation room in Pittsburgh, its consumer affairs office in Winston-Salem, North Carolina, and the eighth-floor conference room at the airline's headquarters in Arlington, Virginia—the place that would come to be known as the Next-of-Kin Room.

Within an hour after the crash, about twenty-five grim-faced managers and vice presidents began to assemble in the big room. A technician hooked up telephones around the table and plugged in a computer that would be used to compile a master list. Flip charts were tacked to the walls so everyone could see important phone numbers and the names of the passengers. A TV in the corner was tuned to CNN.

ANCHOR LINDEN SOLES: I'm going to bring back Leo Janssens, who is the president of the Aviation Safety Institute. As I mentioned earlier, it's a non-profit consumer watchdog group. Mr. Janssens, with the crashes and the run of bad luck that you were mentioning that USAir has encountered over the past five years—this is their fifth fatal crash—in three of those crashes, the aircraft were Boeing 737s. Is there any safety suspicion that we should be reading into that number?

JANSSENS: I really don't believe so, because the Boeing 737 has been in service, airline service I'm talking about, for approximately 30 years. I don't know the exact number of flight hours, but it's got an excellent safety record. Sure there have been crashes, but I ride [the plane] all the time myself. It's just really too early to tell what has happened and therefore I caution people not to be overly concerned at this point about the Boeing 737. USAir normally runs a very good airline. Of course, their safety record over the past five years has been less than admirable in terms of the rest of the industry.

Everything in the Next-of-Kin Room was battleship gray—the walls, the table, even the chairs. The color fit the mood. The USAir employees in the room had all volunteered for this duty, but it was the worst assignment they would ever get. They had to review the reservation lists and tickets for Flight 427, determine who had actually gotten on the plane, and then deliver the horrible news to the passengers' families.

There was no legal requirement that an airline undertake this unpleasant task. After other sudden fatalities, such as car crashes or shootings, local police departments usually did the notification. They sent an officer or a chaplain to deliver the grim news in person. But when a plane crashed, one hundred to two hundred people were killed instantly, and only the airline readily knew their identities. With such an immediate need to inform so many people, it was impractical to alert police in the hometown of each victim. So it had become customary for airlines to deliver the news by phone.

It wasn't fast enough, however. When you're waiting to hear whether someone you love has died, any wait is too long. Television created unrealistic expectations. If the TV networks could cover crashes so quickly, it seemed reasonable to think that airlines could rapidly figure out who was on the plane.

But compiling a list of who actually boarded a plane was surprisingly hard. Many people made reservations and never showed up. Names got misspelled. First and last names got transposed. Long names got cut off by the limits of reservation computers. Babies didn't need a ticket and often were not included on the passenger manifest. Occasionally people from other flights got on the wrong plane and didn't realize it until they were in the air. There was an additional wrinkle: In 1994 the government had not yet begun requiring passengers to show photo ID, and people often traveled using someone else's ticket.

Calls had already begun pouring in to USAir's eleven reservation centers from friends and relatives who urgently wanted to know if their husbands, wives, brothers, sisters, or coworkers had been on Flight 427. The USAir agents could say if other flights had landed safely, but they had no information on 427. They could only promise to call back.

Ralph Miller, a USAir facilities manager and the office computer whiz, was in charge of the passenger list. It was his job to call the airline's Pittsburgh situation room and the Chicago gate agents and go through the list person by person, comparing reservations with the actual tickets that had been collected at Gate F6 at O'Hare.

It was a slow process. The names weren't alphabetized. Miller wasn't sure if there were 125 or 126 passengers. There was confusion about five or six of them, including a two-year-old girl who was sitting with her mother and did not have a ticket. Several Department of Energy employees had been booked on later flights but were allowed to use their tickets on 427. The reservation and ticket totals didn't match. Five or six people who turned in tickets at the gate were not on the reservation list. Another five or six were on the reservation list but had not turned in tickets. Names didn't match. Joan's credit card still had her maiden name, Lahart, so there was confusion about whether the person named on the card and Joan Van Bortel were two different people.

As Miller discussed the last few names for the list, he began to worry. Would he get the list right? Would he miss somebody? Would he put someone on the list who had not been on the plane?

Brett numbly walked outside to his car phone, intending to use it to keep calling the hotel and the airline. That would keep the house phone free in case USAir or Bob Henninger, the man Joan was meeting in Pittsburgh, called back. But as the night wore on, Brett became increasingly convinced that Joan had been on the plane.

When Henninger finally did call, Brett's friend Craig Wheatley answered the phone. Henninger said he had gone to the Pittsburgh airport to meet Joan. At first, the flight was listed as fifteen minutes late. Then it was deleted from the TV monitors. When he went to the front counter, he was told that the plane had crashed.

Craig hung up the phone and came outside to tell Brett. He was a big burly guy who didn't usually show emotions, but now he was shaking his head, crying.

He said, "I'm sorry, man."

Brett just stood there, stunned. He felt like he was melting, like his shoulders could not bear the weight. At some point he wandered into his bedroom and lay on the bed on his stomach. He cried so hard that the tears streamed down his face and off his chin.

Brett's parents, Bonnie and James Van Bortel, drove to his house and stayed with him as he kept dialing USAir on the car phone, trying to get confirmation that Joan was on the plane. It had been four hours since the crash, and the airline still couldn't say if she had been killed.

"I need confirmation!" Brett told his mother.

"You know Joan would have gotten in touch with you if she was okay," his mother said. "She's gone."

But Brett kept calling. When he finally got through, he screamed at the USAir employee, "Goddamn it! My wife is dead and you can't tell me anything!"

"Hold on, please," the USAir employee said.

Minutes went by. When the man finally came back on the line, he said, "We don't have anything at this time. We'll try and let you know as soon as possible."

In the Next-of-Kin Room, the USAir managers crowded around the TV every time CNN issued a new bulletin about the crash.

JIM DEXTER, CNN CORRESPONDENT: USAir Flight 427 from Chicago was just about to land in Pittsburgh before continuing on to West Palm Beach, Florida.

FIRST WITNESS: I looked up and I seen a plane. I didn't hear any sound with it and it started nose-diving. And it seemed like it was going to pull up a little bit and it went on one side of its wing and it went straight down into the ground and blew up.

SECOND WITNESS: There was another couple with me and they said, "Oh my God, there's a plane." And we looked up and it looked like, you know, it was smoking and stuff and it just come down and exploded.

THIRD WITNESS: As soon as it went down I seen a big puff of smoke come up and like, sparks and fire.

DEXTER: The Boeing 737 went down seven miles from the Pittsburgh Airport in a wooded area behind a shopping center.

FOURTH WITNESS: Well, the three of us got in the truck and we ran up there in the truck and the third driveway, I think it was, we turned to the right. We must have walked maybe fifty yards and we kept hollering, the plane was exploding, and we kept hollering, "Anybody alive?" because we seen bodies all over the place.

FIFTH WITNESS: Couldn't find anybody, didn't hear nothing. Parts of the plane were laying all over the place. Little fires here and there. It was a bad scene.

When the bulletins ended, the USAir employees shook their heads in disbelief. Why them? They had just been through this ordeal two months earlier with the Charlotte crash. Why again?

The twenty-five phones in the room continued to ring with calls that had been forwarded from the airline's reservation centers. The callers were crying and shouting, demanding to know who was on the plane. But the managers and vice presidents in the room were not allowed to say. USAir president Seth Schofield had insisted that no one be notified until the list was complete. Even if Ralph Miller had confirmed that Joan was on the plane when Brett called, the USAir managers who were answering the phones were not allowed to tell him. They could only take messages and place them in a box, where they were sorted by passenger name and prioritized so immediate family members would be called back first.

USAir was in chaos. The company had more experience dealing with crashes than any other airline in the 1990s—five in five years!—and yet it was overwhelmed.

There were communication foul-ups between the airline's eleven reservation centers and the Next-of-Kin Room. Some family members were given the direct phone number to the room, others were not. Some USAir employees in the room had experience working on past crashes, but many others didn't. And none of them had any formal training about what to do or what to say.

Each employee in the room was assigned about seven victims. The employees marked a manila folder for each one and began to fill the folders with reservation records and phone messages from relatives.

Posters were taped to the walls with the names of the passengers. Posters from previous crashes had a line beneath each name so the USAir employees could record where the person was hospitalized and what his or her status was—"critical" or "stable" or whatever. But the status lines were blank for the Flight 427 passengers because they were all dead.

About 10:30 P.M., three and a half hours after the crash, Miller finally nailed down the names of the last few people on the plane. He now had a complete list of the people who had been on Flight 427, but he couldn't do anything with it. Schofield had arranged a quick charter flight to Pittsburgh, but he'd ordered that no families be notified until he approved the list. Now he was en route and could not be reached. None of the sullen-faced executives in the conference room wanted to override their boss. And so the people in the Next-of-Kin Room could only sit and field angry calls, without saying what they knew.

Finally Schofield landed in Pittsburgh, reviewed the list, and gave the go-ahead for the calls to begin. It was about midnight now, five hours after the crash.

"We're handing out a confirmed list," Miller told the group. "Throw anything else away. If you get calls, you can find out the next of kin and notify them."

The managers in the gray room had a script that went something like this: "This is _____ from USAir. I'm sorry to confirm to you that _____ was on board Flight 427 and all passengers are presumed to have died."

Some of the employees retreated to private offices so they could be alone when they delivered the news. They took frequent breaks, walking around the deserted hallways of the USAir legal department.

In the jargon of the airline industry, the count of passengers and crew on a plane is known as "souls on board," or SOBS. It refers to the complete count of crew and passengers, to eliminate confusion of whether crew members were included. The USAir managers now had to deliver the horrible news about the souls on Flight 427.

Brett had ended up at his parents' house, awaiting the official word that his wife was dead. He drank a glass of red wine and then fell asleep on the couch. When he woke up, there was a brief moment when everything seemed okay. Then it hit him. The plane crash. Joan was gone.

USAir had tried to reach Brett throughout the night at his house on Riedy Road. When the callers had trouble getting him, they apparently contacted the police department in Lisle to make sure he was okay. About 7 A.M., USAir finally tracked him down at his parents' house.

"Mr. Van Bortel," the airline representative said, "this is absolutely confirmed, sir. Your wife was on the plane last night." The USAir guy sounded weird, almost excited about it, like an announcer telling Brett he had just won a sweepstakes.

Flowers began to arrive. Friends started dropping by to console him, bringing big trays of cold cuts and baked goods. After a while, Brett felt as though the walls were closing in. He took a walk across the road to a nature preserve with a close friend who had been one of Joan's bridesmaids. It seemed to him as if days had passed since the crash, but it had not even been a full day yet.

A woman from USAir called and said she would be Brett's family coordinator. She asked what Joan looked like and what clothes, shoes, and jewelry she was wearing. Brett thought the jewelry might provide some clues, especially since her engagement ring was one of a kind. The woman also asked him to send dental records to help identify Joan's body. When Brett called the dentist to ask for the records, the magnitude of the devastation struck him. There was no body.

The flowers kept coming, filling every room in his parents' house. Brett needed to get out again, so he went for a run in the forest preserve. He and Joan had often hiked through the preserve and played touch football there with friends. He ran a five-mile loop, cut through the woods, and then sprinted up Mount Trashmore. Up and back, up and back he sprinted, trying to burn off the anger and despair.

He wondered what life would be like without Joan. He had always thought they were meant for each other. He often quoted that old country-western song, that the right woman can make you and the wrong woman can break you. She was the right one for him.

That night he talked to his uncle, who was a pilot, and asked him about the crash and whether the government would figure out what happened.

"The NTSB is the best in the world at what they do," his uncle said. "If it's possible to find out what happened, they will find out."

4

TIN KICKER

The phone rang just as Tom Haueter was sitting down with a bowl of popcorn to watch *The Forbidden Planet*. He loved sci-fi and was a big fan of the film, which set the standard for outer space movies when it was made in 1956. Haueter wasn't supposed to be on call for the NTSB's Go Team on this particular night, but he had switched with another investigator who wanted the week off. It would be Haueter's job to figure out why Flight 427 fell from the sky.

Within minutes he had two phone lines going, discussing arrangements with the FAA and his colleagues at the NTSB. "We've got a bad one," he told his boss Ron Schleede. "USAir just lost a 737. It went off the radar near Pittsburgh."

Haueter's first priority wasn't to solve the mystery, it was to find a hotel. He needed beds for several dozen investigators, a meeting room to serve as a command center, and a room for press conferences. Finding a place was difficult because USAir had snatched all the hotel rooms in the area in the first hour after the crash.

Haueter tried to call USAir's accident coordinator, George Snyder, but kept getting a busy signal. When he finally got through, he persuaded Snyder to relinquish a Holiday Inn near the airport. Haueter then had to arrange for fax machines, copiers, and a dozen extra phone lines, including a special line that was for his use only, so he could receive calls from NTSB headquarters. He also had to worry about coffee. The agency's rules were explicit: It would not pay for coffee. But hotels often provided regular and decaf on the big buffet tables without getting approval and then included the expense on the bill. He told a Holiday Inn employee, "We don't want to see the big coffee bar set up."

Haueter was not a coffee drinker. He had an abundance of energy in his trim six-foot frame and had no need for the extra caffeine. He was always in motion—skiing in Colorado, hanging drywall in his basement, flying his open-cockpit Stearman biplane. The license plate frame on his sturdy old Datsun 280Z read, I'D RATHER BE FLYING.

He had wavy blond hair, a moustache, blue eyes, and skin so fair that he wore a floppy hat when he investigated crashes in the hot sun. In a profession dominated by staid engineers, Haueter was a fresh voice. When he got excited, he was likely to use phrases that came from his boyhood in the small-town Midwest: "Holy mackerel!" "Gee whiz!"

He grew up around airplanes in Enon, Ohio, a one-stoplight town of 2,600 people that was midway between Dayton and Springfield. The town was so small that residents joked it was "none" spelled backward. His father was a prominent helicopter and airplane designer who died when Tom was twelve. After that, Tom spent lots of time with his grandfather, Elmer Vivian Haueter, who introduced him to flying. Tom still has a photo in his office taken on the day he got his pilot's license, showing him as a gawky seventeen-year-old shaking hands with his flight instructor. He named his biplane *E.V.* in honor of his grandfather. He preferred the initials—there was no way he was going to name his plane the *Elmer Vivian*.

He got a degree in aeronautical engineering from Purdue University and then worked as an engineer and consultant for a string of aviation and energy companies. When one eliminated his job in 1984 and offered him a less desirable position, Haueter decided to find something more stable. He took a job at the NTSB reviewing safety recommendations. He was not enthusiastic about being a government bureaucrat, however, and figured he would bail out as soon as something better came along.

Instead, he grew to love the job. He got promoted to accident investigator and enjoyed being a "tin kicker," picking through wreckage of a plane to find what caused the crash. The job got him out the office and gave him a chance to climb mountains, ride in helicopters, and see the world. He also got to put his curiosity to work solving mysteries, figuring out how things worked—or why they didn't.

He discovered that the NTSB was surprisingly powerful. His recommendations to the FAA actually got results. He could look proudly at certain airplanes and know they were safer because of his work. The propeller system in Embraer 120 commuter planes was improved after his investigation found a flaw that caused the 1991 crash that killed Senator John Tower. The landing gear on thousands of Piper airplanes was fixed because Haueter discovered that a crucial bolt was prone to crack.

Seeing those changes was the reward of working for the safety board that didn't show up in any paycheck. Every day Haueter could wake up and muse about how many lives he had saved that day.

He was a closet Trekkie. He didn't dress up like a Klingon or hide a phaser in his underwear drawer, but he enjoyed the way *Star Trek* explored issues like race relations and the hazards of technology. He liked how everyone on the spaceship worked together. The people in the NTSB could learn a thing or two from the crew of the *Enterprise*.

Haueter met his wife, Trisha Dedik, in a carpool. They both lived in Great Falls, Virginia, and commuted thirty minutes to the concrete valley of federal buildings along Independence Avenue in Washington. Dedik, who had been divorced for a few years, liked the fact that Haueter could put aside his career to have fun on weekends. He wasn't married to his job like so many Washington men. They began dating in 1988 and were married in 1993.

Dedik also had a fast-lane government job, as director of export controls and nonproliferation for the U.S. Department of Energy. That meant she was in charge of The List, the countries that were allowed to get nuclear fuel and technology to make bombs. As she put it, her job was "to make sure the Husseins of the world can't get their hands on nuclear weapons." Haueter and Dedik weren't Washington celebrities, but they both had unsung government jobs that made the world safer. Haueter's work led to better airplanes. Dedik's kept the world from getting nuked.

As one of the rotating Go Team leaders, Haueter had grown accustomed to wearing a pager, carrying a cellular phone, and being called at home in the middle of the night. His Go Bag was perpetually packed with the tools of an accident investigator—an NTSB baseball cap, a first aid kit, gloves, and government forms. An avid juggler, he often took along a set of juggling balls to relieve his stress—although he forgot to pack them for this trip.

Haueter was a mechanical wizard who loved solving mysteries big and small. On one of his early dates with Dedik, he waited in her kitchen as she was upstairs getting ready. When she came down, her kitchen faucet was lying in pieces in the sink and Haueter was examining the inner workings.

"What are you doing?" she asked.

"I just wanted to figure out how it worked," Haueter said.

He could build or fix practically anything. He built the interior walls in his basement and transformed a bare patch of concrete into a fancy bathroom. He often overbuilt, using an extra two-by-four when one would suffice. "Don't give him a project you ever want to take apart," Dedik said.

When he bought a vintage Stearman biplane in 1984, it arrived as a pile of rubble. For six years, he painstakingly reassembled the plane, replacing the rotten wing spars, covering the wings and fuselage with fabric and stitching it together with a special needle and thread. The result was a spectacular aircraft that he took for weekend hops around the Virginia countryside.

His bosses considered him one of their best investigators. He was a smart engineer who understood an airplane's complex systems, a cautious detective who did not jump to conclusions, and a good manager who could deal with the egos involved in a big investigation. "If two 747s collide over New York," said Schleede, "Tom can do it." The only complaint that the top NTSB officials had about Haueter was that he could sometimes be too nice. He needed "to bare his teeth a little more," Schleede said.

Haueter was forty-two but still showed a trace of the gawky teenager in the photograph of his first solo. His boyish looks and friendly demeanor occasionally made people question whether he was in charge. A Continental Airlines pilot once balked at Haueter's request to ship a flight recorder, even after he flashed his NTSB badge. At crash sites, Haueter often wore a shirt and tie so people would realize that he was in charge—in contrast to other investigators, who wore their NTSB jumpsuits.

The problem had bugged Haueter for years. He felt the old tough-guy approach of running an investigation wasn't effective anymore. You had to be open to suggestions and new ideas. Employees needed to feel free to express their thoughts. Yet he occasionally felt out of step at the safety board, which had the macho air of a men's locker room. Any guy who was prone to use "Holy mackerel!" as an expletive had to prove himself.

By the time Haueter and the FAA were ready to dispatch investigators to Pittsburgh, the last airline flights had already departed. The pilots of the FAA's Gulfstream jet, which was frequently used by the safety board, had run out of flying time for the day and needed a mandatory night of rest. So the NTSB and FAA officials agreed to wait until early the next morning.

Haueter got about two hours of sleep, scarfed down a granola bar and a glass of orange juice, and then drove his old VW station wagon to Hangar 6 at Washington National Airport, where the FAA and the Coast Guard kept their planes. As the team members from the FAA and the NTSB began arriving in the hangar lounge, Haueter could see that he was going to have more people than the plane had seats. He asked Ed Kittel, the FAA's bomb expert, if he would take a commercial flight. Kittel agreed, and everyone else piled their stuff in the plane and climbed inside.

The passengers included NTSB chairman Carl Vogt, one of the five political appointees who ran the agency and voted on the probable cause of each accident. The board members took turns on Go Team rotation and led the nightly press briefings at crash sites.

Also on board was Greg Phillips, a frizzy-haired engineer. No one at the NTSB knew more about 737s than Phillips. He had worked with Haueter on a Copa Airlines 737 crash in Panama in 1992 and had spent months analyzing the rudder system of one that crashed in Colorado Springs in 1991. He had kept close tabs on 737 problems ever since. He maintained a list of suspicious incidents in his file drawer, like a detective tracking a killer.

As the FAA jet rumbled through the sky toward Pittsburgh, Haueter and several other Go Team members sat at a conference table in the back and discussed what they knew about the crash. Haueter flipped through the NTSB's report on the Colorado Springs accident and read the board's previous safety recommendations for the 737. He told Vogt about the problems making arrangements in Pittsburgh and the difficulty getting rooms from USAir.

"Carl, when we get there it will be complete chaos," Haueter said. "But don't assume I'm fucking up on the first day. It will get better."

When the plane landed in Pittsburgh, FAA employees were waiting at the airport with the flight and voice recorders found in the wreckage. After the FAA jet was refueled, the recorders were flown back to Washington, where NTSB lab employees were waiting.

The line of rental cars carrying the Go Team snaked out of the Pittsburgh airport and along Route 60 toward Hopewell Township, a hilly suburb about ten miles away. It was 7:30 A.M., twelve hours since the crash. Haueter wanted a look at the site before his first meeting with people from Boeing, USAir, the Air Line Pilots Association (ALPA), and other groups that would be participating.

Crash investigations are like political campaigns—they throw together a diverse group of people for a few weeks of twelve-hour days under extreme pressure. Everything has a temporary feel because so much of the manpower and equipment is borrowed. So it didn't seem odd to Haueter that his first stop was at the showroom of a Chevy dealer, which was being used as a command post for the local emergency response. He introduced himself to the Hopewell Township officials and then accepted a ride up the hill in a Jeep Cherokee.

The sunny weather from the previous day had given way to a thick morning fog. A creepy mist rose from the asphalt. The woods along Green Garden Road were usually a popular place to see deer, but the animals had been scared away by the crash and the invasion of rescuers. Police and state troopers who guarded the site overnight heard pieces of wreckage falling from trees, but there were no sounds of life.

As his Jeep Cherokee climbed a driveway from Green Garden Road, Haueter noticed pieces of airplane insulation in the trees. The team climbed out of the Cherokee and walked into the woods. They saw more wreckage and the first body parts. Haueter saw a leg bone hanging in a tree. He stepped around a wing panel and glanced up. A dismembered arm was hanging from a branch, a wedding ring on one of the fingers.

He walked carefully around the edge of the debris. "Take a look," he told the group, "but don't move anything."

The first goal in every crash investigation is to find the plane's "four corners"—the nose, wingtips, and tail. If they are found miles apart, it means the plane broke up in the air and then rained to the ground, which suggests an explosion or sudden decompression. But if the pieces are all together, it means the plane was largely intact when it hit the ground. As he walked around the site, Haueter saw all four corners. They were horribly mangled, especially the nose of the plane, and he knew it was possible that other parts had broken off the aircraft before it crashed. But so far, the wreckage told him that the plane did not break up until it struck the hill.

Nobody spoke as they absorbed the horror. The woods now had a slight aroma of jet fuel—the plane apparently *did* have fuel on board when it crashed—and the stench of burned flesh. Haueter found that crash sites had unforgettable smells, slightly sweet and sickening. The investigators looked at the spot on the dirt road where the plane had apparently hit and then walked around to see the debris scattered in the woods. Surely this couldn't be everything from the fifty-ton splane. Somebody asked, "Where's the airplane?"

"It's here," said NTSB engine expert Jerome Frechette. "It's all around us."

Most federal agencies decorate their lobbies with color photos of their leaders or tacky paintings from a starving artists' sale. But the NTSB lobby was different. Color photos of burning planes and twisted trains covered the walls. Airline and railroad executives cringed when they saw their mangled planes on public display, but the pictures were a perfect illustration of the NTSB's job: to determine the probable cause of an accident and recommend changes so it would not happen again.

The agency's roots went back to 1908, when the nation had its first fatal plane crash. The army was testing a Wright brothers' plane at Fort Myer, Virginia, just outside Washington, D.C. Orville Wright had offered to take Lieutenant Thomas E. Selfridge of the Army Signal Corps for a demonstration ride. Selfridge was thrilled to get a chance to fly, waving merrily to friends on the ground as the plane circled the Fort Myer parade grounds. The plane was finishing its third loop when Orville heard two thumps. The plane lurched and plunged seventy-five feet into the field. "Instantly the dust arose in a yellow, choking cloud that spread a dull pall over the great white man-made bird that had dashed to its death," the *New York Times* reported the next day. Selfridge was killed and Orville was seriously injured.

When Wilbur Wright first heard about the crash, he was sure that his reckless brother had been at fault. But after he and Orville analyzed the wreckage, they found it was a mechanical problem. The propeller cracked and cut through a wire that held the tail in place, which caused Orville to lose control. Their investigation was remarkably advanced for 1908, uncovering mistakes that they had made months earlier in stress tests for a bolt on the propeller. Wilbur's explanation of the crash was quite similar to the NTSB probable cause statements ninety years later: "The splitting of the propeller was the occasion of the accident; the uncontrollability of the tail was the cause."

As airlines began carrying mail and passengers in the late 1920s, a branch of the Commerce Department was given the responsibility for regulating aviation and investigating crashes. It was a risky time. Of the 268 airplanes in domestic airline service in 1928, about one-third were in accidents.

The government investigated all major crashes, but those involving famous people had more hoopla and greater significance. The 1931 crash that killed Knute Rockne, the legendary Notre Dame football coach, was especially important in establishing the cautious approach that the NTSB uses today.

The trimotored Fokker F-10A carrying Rockne was flying from Kansas City to Wichita when witnesses saw a wing break off. The plane crashed on a farm, killing Rockne and the seven other people on board.

Because the nation was eager to hear how Rockne had died, the Commerce Department's Aeronautics Branch scrambled to tell what had happened. That was a dramatic change for the agency, which had always been secretive about its investigations.

Investigators initially blamed the crash on pilot error. They said the pilot had pulled out of a dive too sharply, which put too much strain on the wing. But then they found an engine with a missing propeller. They reversed themselves and said that ice had come off the plane and broken a propeller blade, causing severe vibration that snapped the wing. Five days later, the investigators changed their minds yet again when they found the missing propeller in one piece. They said the ice had "rendered inoperative certain of [the plane's] instruments" and caused the plane to go into a steep descent. The wing snapped off as the plane came out of the dive. The embarrassing flip-flops prompted the *New York Times* to question in an April 9, 1931, editorial whether accident investigators could truly find the cause: "Who can tell from a mass of tangled wreckage what actually occurred?"

But eventually investigators found yet another cause: structural failure in the wings. The discovery was especially tragic because the Aeronautics Branch had known of the problem with Fokker planes before the crash and was considering grounding them.

The lessons of the Rockne investigation can be seen in the methodical, cautious approach that the NTSB uses today. The board is usually open about what it finds, describing each discovery at nightly press briefings, but investigators are careful never to speculate publicly about the cause of a crash. The probable cause is not announced until the five board members vote, about one year later.

Despite the embarrassing mistakes on the Rockne crash, accident reports from the 1930s show that investigators were becoming better at using wreckage, pilot interviews, and witness reports to determine what had happened. With no radar records to track a plane's flight path, they often relied on witnesses from different towns to create a map of a plane's final minutes.

Investigators of the thirties used the same basic techniques with wreckage that are used today. When they saw lots of pieces spread over a large area, they knew the plane broke apart in flight. Bent propeller blades told them the engine was operating normally when the plane hit the ground. An open drain valve in an empty gasoline tank meant the plane ran out of fuel. Metallurgists learned to distinguish between parts that broke off in flight because of vibration and those that broke on impact.

As planes got more sophisticated, so did crash investigations. X-ray machines were used to find metal fatigue. Investigators began reassembling wreckage to look for patterns in the broken metal. They even used passenger autopsies to solve cases. Flight data recorders got their start in the 1950s, providing basic information about altitude, airspeed, heading, and vertical acceleration on foil strips. If the recorders managed to survive a crash—and many did not—they could give investigators a rough idea of what had happened. Investigators began enlisting help from airlines, unions, and manufacturers to provide technical expertise, an approach that became known as the party system.

Tension has always existed between accident investigators and the agencies that regulate aviation. The Civil Aeronautics Board, which investigated crashes in the 1950s and 1960s, often got into spats with the Federal Aviation Agency (which became the Federal Aviation Administration in 1967). FAA administrator Elwood R. Quesada frequently angered the investigators by showing up at crash sites and spouting theories about what caused the accident. His behavior violated the post-Rockne rules about not speculating in public. The CAB frequently criticized the FAA for lapses in safety, but FAA officials saw that as a self-serving effort by the watchdog to get more money from Congress.

The NTSB was created in 1966 to consolidate the government's safety offices. It investigates all types of transportation accidents—aviation, railroad, highway, marine, and pipeline. The 1966 law says the board should find the "cause or probable cause of major transportation accidents and disasters." That phrase, which dates back to the 1930s, when the Commerce Department was conducting investigations, gives the NTSB some important wiggle room. It is *probable* cause because Congress believed there would be times when no one would know the absolute truth about why a plane crashed.

In a city of bureaucratic elephants, the NTSB is a mouse. It has only 450 employees and has to mooch off other government agencies in virtually every investigation. It calls the navy when it needs divers, the FBI when it needs bomb experts, and the Armed Forces Institute of Pathology (AFIP) when it needs a coroner.

The NTSB is governed by five political appointees who serve five-year terms. They act as judge and jury after a crash. Investigators such as Haueter are the prosecutors who must convince the board that there is sufficient evidence for the probable cause. But the board often modifies or even rejects the staff recommendation.

The board has little official power, but it still manages to have an impact. When the board determines the cause of a crash, it sends recommendations to the FAA, the airlines, airplane manufacturers, and others. The board may ask for new pilot procedures, changes in training, modifications to an airplane, or all three. The targets of these missives often roll their eyes and sometimes refer to the NTSB as the "Not Too Smart Boys." But the recommendations get results. More than 80 percent are enacted.

With such a tiny budget, the NTSB has to rely on the party system to get help for its investigations. The parties—airlines, airplane manufacturers, the FAA, and unions representing the pilots and mechanics—are invited to provide expertise. They work side by side with the investigators. Pilots explain sounds on the cockpit voice recorder, FAA officials explain how they tested a plane, and airline mechanics identify wreckage.

Representatives from the parties become the NTSB posse. They help at the crash site, attend the nightly meetings, and are invited to submit their ideas about the crash. The parties also benefit from being part of the team. If the NTSB discovers something wrong with an airplane, the team members from Boeing can make sure it gets fixed quickly. Likewise, if the NTSB finds that an airline has a shoddy maintenance program, the airline can correct the problem before it causes another crash.

Critics say the party system is a dangerous way to run an investigation. They say big companies such as Boeing and the airlines are more interested in protecting themselves from lawsuits and costly safety fixes than in finding the truth. The critics say the companies can overpower the NTSB and divert attention from the true cause. They liken it to a police homicide investigation. If the party system were used after a homicide, the suspected killer would be allowed to work side by side with the police. He would be given access to all the evidence and allowed to steer the detectives to other suspects. Arthur Wolk, a Philadelphia lawyer who represents families of plane crash victims, said once on *Larry King Live* that the NTSB protects big companies. "We all know that government is nothing more than a vehicle for special interests and Boeing is one of the biggest special interests in this country."

Haueter liked the party system. Yes, it could be loud and ugly. Each party had its own interests at stake. The pilots union often protected its members, while Boeing defended its planes. The parties often clashed like Republicans and Democrats. But the safety board was well aware of their biases, and investigators were smart enough not to be bamboozled. Haueter felt the system provided healthy checks and balances.

Arguments were a big part of the NTSB culture. Although board members and investigators were careful not to speculate publicly about a crash, there were intense debates behind the locked glass doors at the NTSB offices in L'Enfant Plaza. Rudy Kapustin, a former investigator in charge, remembers having frequent loud arguments with colleague Bud Laynor during their daily commute home to the Maryland suburbs. They would argue nonstop during the thirty-minute drive and then exchange friendly good-byes. When they drove to work the next morning, they would pick up the argument right where they had left off. Two other safety board employees once got into such a heated argument in a conference room that they started slugging each other and had to be pulled apart. "If people watched the way we worked, they would be totally shocked," Haueter said. "There's yelling and screaming, but it works. All these major issues come out and they all get addressed."

The safety board had a near-perfect record at finding the cause of a crash. After all, aviation was a black-and-white science. Engineers knew the exact speed at which a plane lifted off the runway and when it would stall and tumble to the ground. With flight data recorders and cockpit voice recorders, they could do amazing calculations to figure out why a plane crashed. In the NTSB's thirty-year history, only four major accidents had been unsolved. But one of those—a crash in Colorado Springs in 1991—involved the same type of plane used for Flight 427.

A Boeing 737.

After surveying the horror at the scene, Haueter drove to a USAir office building about ten miles away for his first meeting with the parties and the safety board employees who were taking part in the investigation. He didn't like the fact that the meeting was held at USAir. He wanted an impartial setting, but the hotel meeting rooms were not yet available and he needed to get started.

Representatives of the NTSB, Boeing, ALPA, USAir, the FAA, and the machinists and flight attendants unions crowded into the conference room as Haueter explained the rules and how he was organizing the investigation. The place was so packed that people sat on the floor, stood around the back of the room, and crowded in the doorway. The parties stuck together, like teams getting ready for a big game. In one clump sat ALPA, in another sat Boeing. The conference room had been stripped of any evidence that it belonged to USAir. The walls were bare.

First Haueter explained how the investigation would be organized. Groups with representatives from the parties would look into different factors that may have played a role in the crash: weather, air traffic control, and operations, which covered such areas as fueling and cargo. Other groups would study the plane—its structure, engines, maintenance records, and the systems that moved the flight controls. Additional groups would interview witnesses, listen to the cockpit voice recording, analyze the flight data recorder, and study the pilots, even reviewing the details of their lives for the few days that immediately preceded the crash to see if they showed any signs of fatigue or depression. The group would even track down what Emmett and Germano ate for dinner the night before they died.

Next Haueter went over the rules. The parties were to provide technical help. They would also be in a position to respond quickly if the investigation uncovered a safety problem that needed an immediate remedy. But they could not discuss the accident publicly or talk with the press. He warned them even to be careful what they said if they were out for dinner. "I don't want to hear from Mary, the waitress at Bob's Bar, what the NTSB thinks the cause is," he said.

Haueter, who was sensitive to complaints that he looked young, was pleased that he got an opportunity to assert himself and show he was in charge at a progress meeting that night. As he went around the room asking everyone his or her role in the investigation, USAir vice president Bruce Aubin responded that he was "observing."

"Please leave," Haueter told him.

"I'm not going," Aubin said.

"Yes, you are," Haueter said.

"Our company rules require that a senior . . ."

"No," Haueter said. "Your company rules are in conflict with my rules. Please leave right now."

5

THE FIRST CLUES

The morning after the crash, the two "black boxes" from Flight 427 arrived at the NTSB's laboratories in Washington, D.C. The boxes weren't really black, they were bright orange, but they had earned the nickname because of their mystique. They survived accidents that humans could not, allowing investigators to hear the voices of the dead.

The boxes survive partly because they are in the plane's tail, the section of the aircraft that usually has the least damage. They also are extraordinarily strong, resembling steel toolboxes from Home Depot painted with the words FLIGHT RECORDER DO NOT OPEN. Inside are steel cocoons to protect the audiotape or computer chips. They are built to withstand an impact of 3,400 Gs and a 2,000-degree-Fahrenheit fire for thirty minutes.

The cockpit voice recorder, or CVR, runs a continuous-loop tape of the last thirty minutes before a crash. The tape from the USAir plane had four channels of sounds—one from a microphone in the cockpit ceiling, one from each of the pilots' headsets, and one from an oxygen mask in the jump seat. The CVR tapes are tightly controlled. Transcripts are released to the public, but only investigators are allowed to hear the actual recordings. The only exceptions are the rare instances when the tapes are played in court.

The tapes are creepy, like a cross between a horror movie and the Nixon White House recordings. They allow the listeners to eavesdrop on people going through their daily routine. Pilots talk about the "cabbage patch" (the airline's headquarters) and "putting down the girls." (Pilots still refer to flight attendants as girls. "Putting down the girls" refers to the point during final approach when pilots ask flight attendants to be seated.) They say things in a cockpit that they would never say in front of paying customers—they talk about turbulence so rough they're afraid passengers will start vomiting and they make wisecracks about urinating. More than three-fourths of pilots are heard whistling or singing on CVRs. (Bob Rudich, the father of cockpit tape analysis, once wrote an article titled "Beware the Whistler," contending that the whistling was a sign of complacency.) They chat about birds, food, weather, union work rules, and football scores.

The tapes can be embarrassing to airlines, revealing amazing sloppiness in the cockpit. Pilots of a commuter plane on a training flight in Nebraska, for example, sounded like teenagers out for a joyride. "Ye bo, look at all those softball fields. I can really groove on them," one pilot said. "We're just like cruisin' along here, aren't we? We're just, like, toolin'." They talked about trucks and a prank where one pilot had used a Mr. Potato Head as a hood ornament. The captain said he wanted to pull another prank, using a front-end loader to place a friend's Jeep on top of a fuel truck. A few minutes later, they apparently tried to execute a barrel roll and crashed in a field.

The pilots on Eastern Airlines Flight 212 in 1974 made racial slurs and gabbed about politics. "Well, hell, the Democrats, I don't know who in the hell they're going to run," the first officer said. "If they're going to run Kennedy, that's . . . "

"That's suicide," the captain said.

The political gabfest went on for thirty minutes. The pilots talked about busing, the Vietnam War, Arab investments in the United States, President Ford's pardon of Nixon, and the implications of Chappaquiddick on the political future of Senator Edward Kennedy. They were so deep in conversation that they silenced the warning systems on the airplane, ignored standard procedures, lost track of their altitude, and crashed short of the Charlotte, North Carolina, airport.

"Hey!" said the first officer right before impact.

"Goddamn!" said the captain.

When pilots realize they are about to die, their reactions vary. Some plead with the airplane. "Up, baby," begged the captain of an American Airlines jet as it was about to plow into a Colombian mountain. Others are resigned to their fate. "We're dead," said the first officer of a Southern Air Transport cargo plane just before impact. A few shout final messages to their wives and girlfriends. "Amy, I love you!" cried the pilot of an Atlantic Southeast commuter plane just before it hit the ground. Many pilots curse, although the words have changed over the years—they used to nearly always say "shit," but now a growing number say "fuck."

The orange CVR box from the USAir plane was badly mangled, but the tape inside the steel cocoon was fine. In fact, it was one of the clearest the safety board employees had ever heard. Both Emmett and Germano had worn "hot" mikes—headsets similar to the ones astronauts wear. The mikes were so close to the pilots' mouths that they picked up every word clearly. They even recorded the pilots' breathing.

The room where the NTSB played tapes of pilots dying was the size of a small bedroom, with a conference table at the center and six chairs around the edge. At the end of the table were a computer and a small mixing board that allowed NTSB technicians to isolate sounds. Cockpit posters were tacked to the walls so team members could look at the switches and gauges that the pilots were using.

The job of the voice recorder team, which included representatives from the NTSB, Boeing, ALPA, the FAA, and the other parties, was to compile a transcript of the full thirty-minute tape, from the routine chatter at the beginning to the dramatic final seconds. The tape indicated that the pilots were fighting for control, but they never talked about what was happening. Germano said "Hang on" four times but never said why. Emmett cursed but said little else. The most haunting comment came just as the plane's stall warning sounded, when Germano asked, "What the hell is this?"

Bud Laynor, the NTSB's deputy director of aviation safety, called Haueter in Pittsburgh and told him, "This crew had no idea what happened. They never realized what was going on."

The other orange box, the flight data recorder, takes constant measurements such as altitude, airspeed, and heading, allowing investigators to find out how the plane was behaving shortly before the crash. Primitive recorders were used on the first airplanes. The Wright brothers used one to keep track of airspeed, time, and the engine. Charles Lindbergh had one on the *Spirit of St. Louis* that measured altitude and time, to make sure he didn't cheat on the world's first nonstop flight from New York to Paris.

The government issued its first requirement for planes to have recorders in 1941, but the order was rescinded because of maintenance problems and poor reliability with the early boxes. The pilots union, ALPA, fought against having them on commercial planes because of fears that the recorders would be used as mechanical spies. But finally the union relented, partly because a recorder cleared an ALPA pilot who was falsely accused of flying too low. In 1957 the government issued another mandate that planes be equipped with recorders measuring airspeed, altitude, heading, and vertical acceleration. The boxes were primitive—a stylus moved up and down, scratching continuous lines on a strip of foil—but many of them survived crashes and provided valuable clues.

Today, modern recorders store their data on a durable computer chip that can take hundreds of measurements. It records basic parameters such as airspeed and altitude, and it shows what the pilots were doing—whether they were pushing on the rudder pedals or turning the wheel. That information can be especially valuable because it answers the man-or-machine riddle of many crashes.

The recorder on the USAir plane had only thirteen parameters—altitude, airspeed, heading, pitch (whether the nose was pointing up or down), roll (whether the wings were level or rolling down to the left or right), and engine power. It had only two measurements that told what the pilots were doing in the cockpit. One showed when they were pushing the button to talk with air traffic controllers, which allowed investigators to synchronize the flight data recorder with tapes from the CVR and the Pittsburgh control tower. The other showed whether they were pulling or pushing on the control column, the "stick" that made the plane climb or descend. The recorder did not measure what was happening with the rudder or whether the pilots were pushing on the rudder pedals. Haueter's investigators would have to figure that out by themselves.

The labs of the NTSB are messy places. In the metals lab, twisted pieces of airplane wreckage are spread on a countertop like body parts awaiting an autopsy. In the flight recorder lab, mangled orange boxes are piled on a table, many still caked with mud. On a nearby wall is a bank of gadgets that look like a dozen VCRS—computers used to download the information from flight recorders. Another computer can convert the data into a color animation, to show a plane crash like a Saturday morning cartoon.

Technicians in the lab could see that the data recorder from the USAir plane was badly damaged. Dirt and yellow insulation from the plane had gotten inside the box when the 737 struck the hill. The steel cocoon that protected the computer chip had broken away from its mounts and smashed the circuit boards inside the recorder. But the cocoon had done its job. The data were fine. The technicians transferred the data into a computer, converted the raw numbers into rows and columns that were easier to read, and zapped it all by modem to the Pittsburgh command center at the Holiday Inn.

The command center had become a chaotic place. The phones rang constantly with calls from witnesses and others with theories about the crash. A swarm of people converged on Haueter every time he walked into the room, bombarding him with questions about computers, meeting times, phone calls, logistical arrangements. He wondered if he would ever get a chance to actually investigate the crash.

The first person at the Holiday Inn to see the data from Flight 427 was John Clark, a white-haired NTSB engineer. He sat cross-legged on the floor in a corner of the room, studying the results on his laptop computer. The numbers showed the plane was descending from an altitude of 5,984 feet when the left wing dipped. The wing stayed down for about fourteen seconds, then started to level off, then rolled down again. The nose had been up slightly when the wing dipped, but the nose quickly plunged toward the ground. The vertical Gs—the forces of gravity on the plane—told a frightening story. As the plane spiraled toward the rocky hill, the centrifugal force on the passengers and crew reached nearly 4 Gs. That meant a two-hundred-pound person like Emmett would have felt like he weighed eight hundred pounds. The numbers showed how quickly the pilots lost control. Just twenty-eight seconds elapsed from the first hint of trouble until impact.

As Clark looked down the column for the plane's heading, he saw something unusual—an abrupt change, which meant the big 737 had suddenly yawed to the left like a car beginning to skid sideways on a wet road. Other measurements showed that a split second later the left wing rolled toward the ground and the plane plunged nose down. Clark knew many things could make a plane yaw and roll like that, but the most likely was a sudden move by the rudder.

He walked over to Haueter. The data were still rough, he said, but the big shift in heading was significant. "There is something going on here with the yaw," he told Haueter. "It looks like this airplane had some type of rudder event."

It was an encouraging lead. But NTSB investigators have an old saying: Never believe anything you hear in the first forty-eight hours. Early clues can be overrun by new evidence. The first few theories about a crash—known affectionately as the *causes du jour*—often don't pan out.

Still, Haueter felt confident that he would be able to solve the mystery. It was only Day 1. They had good data and a clear CVR. They were making progress.

Before the crash, the hill off Green Garden Road had been a peaceful retreat from city life. It was a thirty-minute drive from downtown Pittsburgh, with thick woods separating it from the steady traffic on Route 60. Children often picked wildflowers in the meadows. Deer wandered through the woods and drank from a creek at the bottom of the hill. The only signs of urban life were the USAir jets that flew overhead, making their final approach to the Pittsburgh airport.

Flight 427 crashed on secluded land owned by George and Mildred Pecoraro, who had lived there for nearly thirty years. They had been displaced before. They lost access to the land when Route 60 was built in the early 1970s, but they moved back after they bought right-of-way and built a dirt road in 1981. They lived in a two-story house at the top of the hill, about one-fourth of a mile from where the 737 crashed. The night of the tragedy, they ended up in a nearby Hampton Inn and weren't fazed by the fact that they were assigned to Room 427. Mildred said they weren't superstitious.

Thousands of fragments from the big plane were blown hundreds of yards away, into fields owned by George David, a police officer in nearby Aliquippa. David grew hay on his 61 acres and loved the solitude of the place. The deer were so friendly they would eat apples right out of his hand. But now, the day after the crash, his peaceful hill looked like the site of a military invasion. Yellow and red police tape was strung around the trees. Helicopters pounded overhead as trucks from the National Guard, the Salvation Army, USAir, and Allegheny County brought supplies and volunteers. Tents were set up along the dirt road as field offices for the Beaver County coroner.

Down the hill, the Green Garden Plaza shopping center had become the nerve center for the crash, with TV satellite trucks parked bumper to bumper and more than two hundred reporters crowding around the command center. The Green Garden merchants all pitched in. The Hills department store gave blankets and other supplies. The Chevy dealer became a temporary headquarters for the Hopewell government. Stress debriefing was available at the New York Pizza Shop.

When Haueter saw the body parts scattered around the hill, he decided to treat the site as a biohazard area. Investigators had rarely worried about diseases before, but he had just taken the government's training on biohazards and felt there was enough danger from blood and fluids to justify employing the full OSHA protections. The investigators would have to wear plastic suits.

Several local officials disagreed. Beaver County coroner Wayne Tatalovich told Haueter the plastic suits weren't necessary. Sure, there were lots of fragmented bodies, Tatalovich said, but there wasn't much blood. He felt the site would not be any more dangerous than a morgue. They argued for a while, but Tatalovich wouldn't budge. He said Haueter's team could wear the suits, but the coroners would not. Haueter warned everyone at the first meeting, "This is a biohazard zone. All safety board employees will have to respect that. I can't force anyone else to wear them, but it's a good idea."

When the USAir plane plowed into the gravel road at 300 miles per hour, it shattered like a crystal vase thrown on a concrete driveway. The 109-foot plane splintered into hundreds of thousands of tiny fragments, many no larger than a plane ticket. The USAir logo was usually found on hundreds of items in the plane, but one of the airline's mechanics noticed an odd pattern to the logos he found in the wreckage. They said "US" or "USA" or "Air," but he could not find any logos that were intact. Everywhere he looked, USAir had been torn apart.

The site was littered with seat cushions, hundreds of shoes, and thousands of *BusinessWeeks* with the headline THE GLOBAL INVESTOR on the cover. The magazines were everywhere.

Passengers' belongings were scattered through the woods and on the road. It was as if the crash had taken a snapshot of each person's life, revealing that person through his or her possessions. There were plaid boxer shorts and another pair with red diamonds; sweatshirts from Purdue University and the New York Renaissance Festival; T-shirts for Hooters, Soldier Field, the Chicago Bears, Harley-Davidson, and Bugs Bunny. There were lots of mangled and burned books: *Forrest Gump,* the *Pocket Prayer Book,* Rush Limbaugh's *The Way Things Ought to Be, The Chamber,* by John Grisham, a management training manual called *Firing Up Commitments During Organizational Change,* and a copy of the Bible. Investigators also found lots of everyday stuff: a garage door opener, family snapshots, a Swiss Army knife, pocket calculators, Kodak film, a rosary, and a teddy bear.

While many of the local volunteers were vomiting in the woods or sobbing at the horror, Cox remained unfazed. He had been to many crash sites as an investigator for the pilots union, and he knew the smells and didn't dwell on the body parts. He was there to look at wreckage, not people. He had been appointed the ALPA representative on the systems group, which was shaping up as the most important group in Haueter's investigation. The systems members would examine the plane's flight controls and hydraulics to see if they had caused the crash.

Each morning Cox joined the other members at Green Garden Plaza to be sealed inside his rubber suit. It was like getting dressed for surgery. They had latex gloves, boots, and surgical masks. The boots and gloves had to be taped to the suits, which made the outfits unbearably hot. With everyone wearing an identical white rubber suit, it was also difficult to tell people apart. They eventually wrote their names across their backs, as if they were wearing football jerseys. Everyone involved in the investigation also had to wear a colored bracelet to get access to the site. To foil trespassers, the color changed every day. Several people had been arrested trying to sneak onto the hill to take pictures of the wreckage and the body parts.

Cox's first assignment was to pick through the flattened wreckage of the cockpit. Getting to it was difficult because of the trees and hilly terrain, so the systems group enlisted the help of the Allegheny County Delta Team, a paramilitary group of public works employees who responded to the crash like they were invading Kuwait. "You want a road? We'll build you a road," said one Delta member cheerfully. A few hours later, there was a gravel road straight to the cockpit.

Cox found that picking through the wreckage wasn't easy. The investigators had to use a pulley on a big metal frame to lift the largest pieces. Some were buried several feet underground. Others had been flattened like aluminum cans. Much of the wreckage was buried beneath a thick layer of wire that looked like burned spaghetti. Their first priority was to see what they could learn from the gauges and switches. Cox was especially interested in finding bulbs from the cockpit warning lights. Lightbulb filaments stretch when they get hot, so the investigators could tell if a warning light had been on by measuring the filament. But Cox discovered that every light had been shattered.

The softest things in Ship 513 had survived with the least damage—seat cushions, handbags, and hundreds of shoes looked fine. But the rest of the plane was torn apart and hard to identify.

"That looks like junk," said one of the Delta Team members, pointing to some twisted metal.

"It's a nose-gear strut," said Cox.

The gauges provided a few clues. The captain's airspeed indicator was covered with mud, but when Cox cleaned it off, he saw the needle had stopped at 264 knots—the plane's speed when it hit. A needle on the hydraulic pressure gauge indicated that the B system was at 3,100 pounds, which told Cox that it had full power when the plane crashed. The plane's hydraulics are crucial because they move the landing gear and flight controls such as the rudder, elevator, ailerons, and spoilers. The needle for the second hydraulic system—the A system, which moved the landing gear—was missing.

Elsewhere on the hill, other teams were finding more clues. The 737's engines were badly damaged, but the members of the power plant group could see that the fan blades were bent opposite to the way they rotated. That meant the engines were running at impact, which ruled out the possibility that engine failure had caused the crash. Everyone looked for parts from another plane, on the theory that the big 737 might have collided with a Cessna or a Piper. But so far, none had been found.

Cox was perfect for the systems group. He was a 737 pilot who knew every inch of the cockpit and, like Haueter, he loved dissecting the mechanical and electrical systems that made the plane fly. He was not a do-it-yourselfer as Haueter was—Cox was away from home too much to have time to build things—but they shared a fascination with solving mechanical mysteries.

Cox was a meticulous guy who kept his life and cockpit carefully organized. His bookshelf was a reflection of his personality: All the Tom Clancy hardbacks were on one shelf, all the books about flying on another. Paperbacks were together, separated from the hardbacks. An errant copy of his wife's *Martha Stewart Weddings* put in an appearance on the maritime shelf, but it didn't stay long. Cox kept their finances on their home computer, and he maintained precise records about where their money went. When his wife, Jean, came back from shopping, she had to separate the expenses into categories such as Household and Gifts.

Whether he was flying a difficult approach into O'Hare on a stormy night, driving 100 miles per hour in his sports car, or just balancing his checkbook, Cox was in control. On his business card he was "Captain John Cox." He had 8,000 hours flying 737s, and he loved the plane. He had flown lots of others in his twenty-four-year career, but he had always preferred the 737. He liked its smooth landings and the solid way it handled a crosswind. "The airplane tends to make you look good," he said. He loved the challenge of mastering a machine, maneuvering the fifty-ton bird through winds and clouds and heavy rain and still managing to touch down so gently that the passengers in back could barely feel it.

Cox also craved speed. The speedometer on his fire-engine-red Acura NSX went to 180 miles per hour. He had gotten the needle up to 125. He referred to the $80,000 sports car as "the toy," but he treated it with reverence. When he parked at a store or restaurant, he put it in a remote corner of the lot, parked at an angle across two spaces so no one would nick his doors. He kept a cloth cover on the car, even when it was inside his garage. When he removed the cover, he folded it up carefully, one side at a time, as if he was folding a flag.

The son of a Birmingham banker, he grew up in a family that had no connection to aviation. But he got interested in airplanes as a toddler. One of his first words, uttered when he was two, was "Constellation," the big plane that he watched taking off from the Birmingham airport. He got his private pilot's license at age seventeen, flew charters and corporate planes at eighteen, and then joined Piedmont Airlines at twenty-six. He became a USAir pilot when the two airlines merged in 1989.

Cox was trim, with the graying hair, silver moustache, and tanned good looks that seemed standard issue for an airline pilot. He was one of ALPA's technical experts, a rare pilot who understood the complex engineering of the planes he flew. Even his doodling was intricate. The margins of his notepads were filled with complex geometric figures that looked like M. C. Escher drawings.

His union had a Jekyll-and-Hyde reputation in aviation safety. C. O. Miller, a former NTSB official, often said that ALPA had done more to promote safety in the skies than any group except the federal government. The union had helped design the national air traffic control network and the instrument landing systems that guide planes toward a runway. It played a big role in changing cockpit design to reduce pilot mistakes (the easy-to-read T design on the instrument panel is one of its legacies), and it was a strong proponent of crew resource management, which has improved communication in the cockpit.

But so far as some critics were concerned, ALPA had a reputation as a union that used safety as an excuse to get more money and generous work rules. Najeeb Halaby, a well-respected FAA administrator from the early 1960s, once said there were two ALPAS—the one that made substantial contributions to safety and the one that masked its economic demands "under the guise of safety." The problem, Halaby said, was that he could never be sure which ALPA he was dealing with.

Among accident investigators, ALPA had a reputation for sometimes making excuses for pilots when there was overwhelming evidence that they had screwed up. The union would claim the pilots were influenced by the design of the plane or try to blame air traffic controllers or some mechanical problem. In some crashes, it was obvious that the pilots had made a stupid mistake—they simply forgot to set the flaps for takeoff or they flew into a bad storm. But ALPA would throw up smoke screens and make excuses.

Cox, however, was respected at the NTSB because he was not a strident unionist. He was regarded as one of the Young Turks at ALPA who were more like accident investigators than defenders of the pilot brotherhood. He had taken the highly regarded accident investigation course at the University of Southern California and followed its open-minded approach. "The evidence leads you where it leads you," he often said. If that meant a pilot was at fault, so be it.

There had long been a culture clash between Boeing and ALPA that was rooted in the starkly different styles of engineers and pilots. Boeing engineers existed in a black-and-white world of data. In their view, if you got enough data, you could do anything—build a perfect wing, design a better engine, or fix a faulty part. But they were perplexed by the macho personalities of the pilots who flew their creations. It didn't help that in 1955 test pilot Tex Johnston shocked Boeing's top brass by making a risky barrel roll in a prototype of the 707 in front of thousands of people. Boeing president William Allen was furious that a pilot had endangered the plane and so many people with such a daredevil maneuver. The engineers also resented ALPA's long fight to get a third pilot in the 737 cockpit. The union said the third pilot was needed for safety, but officials at Boeing and the airlines saw it as a blatant attempt to get more people on the payroll.

On the other hand, pilots complained that the Boeing engineers didn't appreciate them. The pilots felt they had a trait that couldn't be measured on any chart: courage. *They*—not some beady-eyed engineer with a slide rule in his pocket—were responsible for hundreds of lives every day. When a pilot shot a tight approach into La Guardia on a snowy night, all the data in the world would not make the wheels touch down safely unless the pilot knew what he was doing. The engineers had no equations that mentioned guts.

In Hopewell, it was clear that Boeing and ALPA were rival teams. They were cordial with each other, but they didn't mix much. Cox found that the Boeing investigators rarely spoke up and always traveled in packs. When everybody else got together for breakfast or dinner, the Boeing guys would go off on their own. Cox jokingly called Boeing a black hole—information went in, but it didn't come out.

6

THE GLOW FROM
THE HILL

In the first few days after the crash, members of the CVR team listened to the cockpit tape many times. The team quickly identified most of the sounds on the tape, such as the snap of a shoulder harness, clicks from the elevator trim wheel, and the *rat-a-tat-tat* of the stickshaker. But there were a few thumps they could not identify. They listened to the sounds hundreds of times but could not recognize them. The thumps were muted and did not seem to originate from the metal fixtures of the cockpit. It was time for some experiments.

On September 11, three days after the crash, members of the CVR team arrived at Washington National Airport and walked to a gate where a silver USAir plane was parked. Their goal was to record a variety of sounds on the plane's CVR to see if they matched the thumps from Flight 427.

Sounds on cockpit tapes were often as valuable as the pilots' words. Investigators could calculate engine thrust from the distinctive hum of a jet engine. They could determine runway speed by counting clicks heard as the plane's nose wheel ran over embedded lights. Sounds could be displayed on a graph like a fingerprint, with squiggly lines representing volume or pitch. By taking a fingerprint of a mysterious sound and comparing it with one from a known sound, investigators could look for a match.

Al Reitan, a voice recorder specialist with the NTSB, came to the airport with Mike Carriker, a Boeing test pilot, and Paul Sturpe, a USAir pilot. The silver plane at the gate was the same model as the accident airplane, a 737-300, with the same type of cockpit voice recorder. They turned on the plane's auxiliary power unit to provide electricity to the CVR and began a series of tests.

They tried to imagine what might have happened in the cockpit to cause the thumps. They flipped switches and yanked on levers. They dropped notebooks on the floor and turned the trim wheel. They fiddled with the clip that held pilot checklists. They pulled on the flap handle and triggered the stickshaker. They stomped their feet in the doorway and in the first-class galley.

Then they returned to the NTSB offices at L'Enfant Plaza and used a computer to draw the fingerprints. The strange thumps from the original Flight 427 tape showed up as dark spikes, like a fingerprint of a burglar. All they needed was a match.

They ran the new tape through the computer. The stomps and slams from the test also showed up on the screen as spikes. But they were distinctly different from the mysterious thumps on Flight 427. The fingerprints did not match.

Brett felt restless and overwhelmed by all the people who had stopped by to offer their condolences. At one point there were thirty or forty people crowded in his parents' house, all with good intentions, but Brett couldn't take it anymore. He needed to get out of there. He wanted to visit the crash site and say a final farewell to Joan.

USAir had said he could probably see the site and that he might be needed to identify Joan's body. So Brett, his mother Bonnie Van Bortel, Joan's brother Dan Lahart, and Brett's friend Craig Wheatley had piled into his mother's Jeep Cherokee and driven to Pittsburgh. Brett couldn't concentrate on the road, so his mom and Craig took over the driving. As they arrived in Hopewell Township, the hill where Joan had died now glowed a brilliant white, lights ringing it like a crown. It almost looked beautiful.

"Craig, can you pull over?" he asked. They stopped about two hundred yards from the exit for Green Garden Road. Brett got out, knelt in the asphalt by the guardrail, and said a long prayer.

He was numb that weekend, still trying to make sense of the fact that Joan was gone. Everywhere he went, he carried a crystal frame with their wedding photograph that he had picked up on the way out of the house. Joan looked beautiful in the photo, with her hair pulled back, a perfect smile, and her hand resting gently on Brett's arm. But the picture called attention to his loss. When he checked into a downtown hotel where USAir had rooms for the families, the bellman saw the picture.

"Did you know someone in the crash?" he asked.

"My wife."

Suddenly Brett was a celebrity, but for all the wrong reasons.

USAir was paying for the hotel rooms, meals, and other expenses. The company had offered to fly Brett and his family to Pittsburgh, but the last thing he wanted to do was fly, especially in a USAir plane. At the hotel, he met the airline employee who was assigned to be his liaison during his stay. A saleswoman for USAir, she seemed poorly trained and unprepared for the job. When they met, she broke down and cried.

Brett told her that he respected anyone who would volunteer for such a difficult job, but over the next two days, he realized that she was clueless. She rode around in a white limo and couldn't remember the hotel where she was staying. She had a cellular phone so Brett could get in touch with her anytime, but the battery was dead and she did not know how to charge it. She kept flipping through legal pads, reciting directions from her bosses, but she was unable to answer his most basic questions if they weren't addressed by the instructions on the pads. She seemed more interested in the airline's needs than in Brett's. The only time she seemed animated was when he mentioned talking to the media. "You can talk to the media," she told him, "but we'd advise against it because you're going through a period of grieving."

She tried to explain how they were going to identify the bodies, using a grid system to locate the body parts. But Brett didn't want to hear about it. "Great," he said. "I'm glad you're getting a little science lesson out of it while there are pieces of my wife laying up in that hillside."

USAir had reversed itself. There would be no visit to the site and no opportunity to look at Joan's body. Instead, USAir asked him to send books and perfume bottles that might have her fingerprints.

Airlines had a long tradition of helping families after a crash. The companies believed it was the compassionate thing to do and also was good for public relations. Most airlines assigned an employee to be a liaison with each family and paid for the family's travel, funeral expenses, and many other costs. The airlines bought meals, made mortgage payments, and occasionally even paid a speeding ticket for a grieving family member.

Critics said there was an ulterior motive for the corporate kindness. The airlines could collect a dossier on the victims that could be used in court to fight for smaller awards. If the airline learned that a victim had a drug problem, for example, it might convince a jury to reduce the amount of the award because the victim would have had a shorter life expectancy. But the airlines insisted that their family coordinators were to help grieving relatives, not to ferret out details about the victim.

Some airlines were better prepared for a crash than others were. They had thick notebooks that spelled out how they should respond minute by minute, and they offered special training for employees who worked with the victim's family. But not USAir. It was caught unprepared for the Hopewell crash, even though it had just handled the Charlotte accident two months earlier and had had three other crashes within the previous five years. No airline had as much experience with crashes in the 1990s as USAir did, but the company still seemed bewildered about what to do. USAir's director of consumer affairs had written a plan to revamp the response for families and establish special training for the airline coordinators, but the plan had not yet been approved by top executives when the Hopewell crash occurred. As a result, Flight 427 families experienced a wide range of responses from the airline. Some said their USAir coordinators were compassionate and organized. Others, like Brett, thought they were ill prepared and insensitive.

To make matters worse, Brett and his family had to deal with the news media. They had been badgered by reporters the day after the crash, but Brett had refused to talk. A TV news crew tailed him as he drove from his parents' house to his home in Lisle and then ran up to him in his yard. He told the reporter to leave. "I can't do this right now," he said. Another reporter was rude when the family declined to talk, but he eventually left.

The reporters were engaged in a painful ritual that follows a tragic death, whether the death results from a plane crash, a car accident, or a tornado. Reporters disliked the practice as much as the families did, but the stories were an expected part of news coverage after any disaster. They put a face on the tragedy. In newsrooms all over the country, editors studied the Flight 427 passenger list for anyone from their area. If they found someone, it gave them a stake in the crash.

A reporter was then assigned to find out everything possible about that victim. The assignment meant checking clips in the newspaper's library to see if the victim had been in the news, looking at land records to see what the victim owned, and researching court records to see if the victim had ever been arrested or sued. Such inquiries might sound insensitive, but those sources were all public records, and they spoke volumes about the victim's life. The reporters used city directories and called neighbors, who usually had nice things to say about the victim. The *Chicago Tribune*'s story was headlined CHICAGOANS MOURN THEIR OWN. It quoted an unnamed relative of Joan's saying that she "loved life. She was only going to be gone for a day. Just a day."

Most reporters dreaded knocking on the door of a widower's house and handled the task with sensitivity. Many families were willing to talk. It was their chance to commemorate the person they loved. If families preferred not to say anything, most reporters left politely. A few, however, were so intent on getting a tearjerker of a story and beating the competition that they were rude. After the ValuJet crash in 1996, TV camera crews hid in the bushes in a hotel parking lot and then jumped out when they spotted a distraught family. After the crash of TWA Flight 800, a reporter for a New York tabloid posed as a relative of a victim to get inside the hotel where the families were staying.

On Sunday, September 11, Brett sat in a front pew at St. Margaret Mary Catholic Church in Moon, Pennsylvania, for a service to honor the victims. One hundred thirty-two candles burned on the altar, one for each person on the plane. Brett's mother gave his hand a comforting squeeze. He clutched the wedding picture to his chest through much of the service, but occasionally tipped it down to look at Joan. At one point, a tear streamed from his cheek and splashed on the picture.

TV cameras were clustered at the back of the church, trying to capture the grief, but Brett didn't give them much thought. Then, midway through the service, the priest encouraged the congregation to greet people sitting nearby. When Brett turned to a woman sitting behind him, she shook his hand but looked away nervously.

When the service ended, a different woman sitting beside Brett introduced herself and said her brother was killed in the crash.

"I'm so sorry," he said, hugging her. "It's so easy to forget about everybody else who lost, too."

"I'll pray for you, if you'll pray for me," she said.

"I will, I will pray for you," Brett said. "I won't forget."

As Brett left, he stopped and did an interview with a TV reporter. But he never spoke to the woman behind him, who had turned away when he went to greet her. The next day, when he picked up the *Pittsburgh Post-Gazette,* he was amazed to see the paper had printed his entire conversation with the sister of the victim. The woman behind Brett had been a reporter, writing down every word, yet she never identified herself.

The next day, Monday, September 12, was Brett's birthday. Days earlier, he and Joan had talked about going out to dinner to celebrate. But now he was in Market Square in downtown Pittsburgh for another memorial service. About five thousand people packed into the square, a big turnout that was a testament to USAir's large presence in the city. Brett sat in one of the front rows, beside a man from the Salvation Army. As they sang hymns, Brett couldn't help but notice the guy's voice. He had the worst singing voice Brett had ever heard. Brett chuckled a little and for just a moment felt a break from the relentless sadness.

Brett had never been much of a churchgoer—he believed you did not have to go to church to have religious faith—but he was comforted by the two services. He believed that God occasionally sent you a sign. During the service, a big jet passed overhead, its shadow racing over the crowd. Brett thought it was a sign that Joan was going to be okay.

Within minutes of the crash, reporters began calling USAir's Arlington headquarters to ask if the airline was unsafe. The company seemed to have all the warning signs. It had been in deep financial trouble, losing $2.5 billion since its merger with Piedmont Airlines in 1989. It was under pressure to cut costs to compete with more efficient airlines that were charging rock-bottom fares. And now it had had its fifth crash in five years.

"For USAir, this is Apocalypse Now," Gerald Myers, author of a book on corporate crises, told the *Charlotte Observer.* "This is more than a slippery slope for them; it's a cliff. They're getting themselves in the same position that Exxon got in with the *Valdez* or A. H. Robins with the Dalkon Shield."

USAir's financial problems had prompted the FAA to beef up inspections two years earlier. But the day after the Flight 427 crash, FAA administrator David Hinson said his agency had not found any serious problems. "We deem [the airline] to be safe," Hinson said. "In fact, this afternoon I will be flying on USAir."

At a press conference in Pittsburgh, USAir chairman Seth Schofield was swamped with questions about the airline's safety record. He said the five crashes were not connected in any way.

"If I thought USAir was an unsafe airline, I would put the entire fleet on the ground until any problems were corrected," Schofield said. (That is the standard response from an airline chief when his company's safety record is challenged. ValuJet president Lewis Jordan used nearly identical words after the 1996 crash in the Everglades.)

In a message posted on company bulletin boards, Schofield warned his employees to be ready for rough times. "In the coming days, you will surely hear and read comments in the media and elsewhere that will offend and hurt you. I encourage you to lean on each other for support and, in doing so, you will strengthen each other."

The *New York Times* asked statisticians if USAir's safety record was worse than those of other airlines. Most agreed that plane crashes were random events that had no connection with a particular airline. But Arnold Barnett, a professor at the Massachusetts Institute of Technology, was quoted as saying, "If you got on a random USAir flight in the 1990s, your chances of being killed are nine times as high as if you got on a flight of any other airline."

The hemorrhage had begun. USAir's bookings fell drastically over the weekend.

The dream of every accident investigator was to find the Golden BB. That was the NTSB nickname for some tiny piece of wreckage that instantly explained why a plane crashed. There were just a few Golden BBs in aviation history—a disk from a jet engine that broke apart and caused the crash of a United Airlines jet in 1989, and a latch on a DC-10 cargo door that caused a Paris crash in 1974. But usually investigators had to be plodding and methodical, eliminating one theory after another until they zeroed in on the real culprit. It typically took several weeks before investigators knew the cause, but the NTSB usually took twelve to eighteen months to officially complete a case.

That wasn't fast enough for the news media. Reporters were ruthlessly competitive and eager for scoops, which meant they couldn't wait until the safety board completed its report. In the Flight 427 case, they began speculating about the cause before the NTSB even got to the scene, calling pilots, trial lawyers, and former safety board members and asking them about previous accidents involving the plane. The *Seattle Times* and the *Dallas Morning News* both pointed out that the 737's rudder system had been under scrutiny because of a possible flaw in a hydraulic valve that could make the rudder go the wrong direction. The NTSB frowned on that kind of speculation, preferring to make its own pronouncements as it discovered the evidence, but reporters were merely doing in public what Haueter and his team were doing behind closed doors.

The hunger for information about a crash dates back to the first aviation fatality involving the Wright brothers' plane. Reporters swarmed onto the field at Fort Myer seconds after the crash and ignored requests to move back until the army sent in soldiers on horses that practically trampled the reporters. As recently as the 1980s, reporters were allowed to visit crash sites, stepping over wreckage and dead bodies. But the proliferation of Action News and Eyewitness News and twenty-four-hour cable news channels led to such intense coverage that crash sites are now quickly roped off and reporters are herded as far away as possible. The NTSB typically gives only one or two briefings a day, a short one in the early afternoon and a more detailed one about 7 or 8 P.M.

The NTSB briefings were like a high-stakes game of Twenty Questions. The NTSB stuck to the facts, explaining the evidence without putting it in context. Reporters had to read between the lines and figure out which tidbits were truly important. The game was especially difficult for new reporters, because the NTSB representatives often spoke in jargon, leaving many journalists dumbfounded about what it meant. A handful of aviation reporters from the major newspapers and TV networks knew how to play the game, but most people at the nightly briefings were local journalists who had never dealt with the safety board.

The NTSB's *Aviation Investigation Manual* warned about the press. It said investigators should be careful about using cellular phones because reporters might intercept the calls. It said the investigator in charge "should be aware that reporters will be looking to him/her for body language or facial expressions during the Member's briefing and should, therefore, maintain as neutral an expression as possible." The manual also gave advice on what to say when reporters asked speculative questions: "Rely on tried and true phrases such as 'That is one of the many things we will be looking at'; 'It is much too early to tell as this point'; 'Right now we are not ruling anything out.'"

The first day in Hopewell, someone had leaked to reporters a copy of the air traffic control transcript. Haueter was angry that it was leaked, because it was not the FAA's official transcript (which would not be done for days, after every voice had been identified), and it might contain errors. But as he expected, the transcript dominated the news coverage, along with the preliminary information he released from the flight data recorder.

The second day, investigators on the hill found suspicious parts from the thrust reversers, the engine doors that open when a plane lands. These devices reverse the jet blast to slow the plane on the runway. If a thrust reverser opened in flight, it would be catastrophic, like slamming on the brakes on one side of a car. The 737's engines had safety locks that were supposed to prevent the reversers from activating until the plane was on the runway, but workers on the hill had found evidence that they might have deployed. Also, the workers could not find key pieces that usually locked the reversers in place. At his nightly meeting with investigators at the Holiday Inn, Haueter recounted the findings but warned them not to jump to conclusions.

"Let's not focus just on this thrust reverser," he said. He was skeptical about the theory because it did not match the flight recorder, which showed the plane's engines at idle just before it plunged from 6,000 feet. Also, the fact that it was the *right* reverser did not match the data that showed the plane rolling to the left. The Boeing investigators urged Haueter not to tell the press because the evidence was incomplete. But Haueter and Vogt decided they should tell reporters everything they had, regardless of whether it was incomplete. If they didn't mention the reversers at the briefing, the news would probably leak out anyway. (The NTSB was notorious for leaking to the press. Peter Goelz, the agency's managing director in the late 1990s, joked that the NTSB's official seal should have an eagle clutching a sieve.)

That night at the briefing, Vogt explained the discovery but cautioned reporters that the reversers could have popped out when the plane struck the ground. The press dove for the story. THE CAUSE? MAYBE THE ENGINE, said a headline in the *St. Petersburg Times*. THRUST REVERSER SUSPECTED IN USAIR JET-LINER CRASH/DEVICE COULD HAVE CAUSED NOSE DIVE, said the *Houston Chronicle*.

But the next day the thrust reverser theory unraveled. Investigators found locks that showed the reverser doors were closed when the plane hit. The first *cause du jour* had been ruled out.

Amid all the chaos, Haueter tried to account for every piece of the plane. Nearly all of the wreckage seemed to be on the hill, but a few pieces were turning up elsewhere. A passenger's business card and some light insulation from the plane were found two and a half miles downwind from the hill. Could that mean the plane had exploded in flight? Clark, an expert on airplane performance, sat down with his laptop computer and launched a program he'd written called WINDFALL. The program used information about wind and the plane's flight path to estimate where wreckage might have fallen. It said that if pieces had fallen from the plane, they would probably be behind the shopping center.

On September 14, an army of more than 150 volunteers, search-and-rescue team members, and NTSB employees gathered in the parking lot at Green Garden Plaza to start a massive search in the triangle-shaped area that WINDFALL had identified. For several hours, the team members crawled through bushes, waded through creeks, and peeked in backyards.

One volunteer, a USAir flight attendant, paddled a rowboat to the middle of a pond to investigate a mysterious object floating on the water, only to find it was insulation from a building. Another volunteer found something that looked like a rocket in someone's backyard. It was rushed back to the FAA's bomb expert, who determined that it was a spare part for a home furnace. Searchers in a helicopter spotted a suspicious panel hanging in a tree and a ground team hurried to find it, but it turned out to be a "DuckTales" kite.

When the search ended, the only items found away from the hill were insulation and light debris that had been carried in the hot plume of smoke. Once again, it looked as if the big jet was intact until it struck the hill.

At his press briefing, Haueter recounted the search. He said USAir employees in Chicago reported nothing unusual about the flight. Engine bolts in the wreckage were cracked, but those cracks probably occurred when the plane hit the ground. The plane's logbook showed no problems with Ship 513. They had not found the Golden BB.

ZIPLOC BAGS

Dave Supplee was accustomed to seeing 737s taken apart. A USAir mechanic on the overnight shift in Tampa, Florida, he could fix anything on a plane—radios, hydraulic pumps, even the cranky APU generators that always seemed to be breaking down. Supplee, thirty-six, was a safety official with his union, the International Association of Machinists, and was on call anytime there was an accident involving a USAir plane. An hour after the Hopewell crash, he got a call from John Goglia, the union's accident coordinator. "Let me warn you now," Goglia said in his thick Boston accent, "this one is not pretty. There is total destruction of the aircraft." Two years earlier, Supplee had worked on another USAir crash, Flight 405 at LaGuardia, but that scene wasn't nearly as gruesome as the one in Hopewell.

Supplee was an ideal member for the NTSB's structures group because he could identify the mangled parts lying on the ground and hanging in the trees. "Yeah, this is your air-conditioning bypass valve," he said. "This is your hydraulic pump." His group painted lines for a grid to keep track of the wreckage, so they could look for patterns in where the items landed.

When he first arrived at the scene, dressed in his white rubber suit, Supplee felt a sudden emptiness, as if all the life had been drained from the area. He tried not to think about the carnage around him. When he saw a hand or a foot lying on the ground, he called the coroner's team over. They tagged it, noted the location with a colored flag, and then put it in a one-gallon Ziploc freezer bag. As the week wore on, the foot-high red and yellow flags sprouted everywhere, like survey markers at a construction site. Initially, red flags were supposed to designate body parts and yellow ones, wreckage. But there were so many body parts that they quickly ran out of red and had to use flags of all colors. It was a strange sight—a rainbow of flags flapping in the wind, an unintended memorial to one of the most gruesome air crashes in U.S. history.

Late one afternoon Supplee was assigned to find the plane's cargo doors. There was a theory that one of them might have blown out in flight, so Haueter wanted to know if they were all on the hill. A coroner's team had been working with Supplee's group, picking up body parts, but the coroners had stopped for the day. Just after the team walked away, Supplee discovered pieces of door trim. Figuring that the rest of the door was buried just below the surface, he and other members of the structures group dug into the rocky soil.

When they pulled up the door, they found a woman's arm partially covered by the navy blue sleeve of a flight attendant's uniform. Right beside the arm was a purse. An ALPA investigator opened the purse and found the woman's passport and wallet. He opened the wallet and saw the cheerful picture of April Lynn Slater, one of the flight attendants. Suddenly death on the hill wasn't anonymous anymore.

Supplee looked around in hopes that someone from the coroner's team was still nearby, but they were all gone. It was late afternoon now and everyone was leaving. Supplee realized he would have to leave the arm on the hill for the night. It bothered him that it would be left behind in the darkness until the coroners retrieved it the next day.

A few hours later at the nightly progress meeting, he broke down crying when he told the investigators from the flight attendants union what he had found. When he returned to his hotel room, Supplee called his mother and told her about the horror of the site. She couldn't understand why he had volunteered for the job. "Why are you there?" she asked. "Why are you putting yourself through this?"

"I just couldn't imagine *not* being here," he said. "I *have* to do this." He felt he was making a contribution to safety. That was the paradox about the whole ordeal. Crashes made flying safer.

Many investigators coped with the horror by building imaginary walls. Instead of looking at the body parts scattered around the site, they focused on the wreckage. Haueter told them, "Concentrate on metal, not on people." Supplee followed that advice but found that he still got upset at the end of the day, when he went back to his hotel room and collapsed on the bed. He would turn on the TV, hoping to forget about the crash, but it would be on every channel.

As the hill got soaked by storms and then baked in the sun, it took on the horrible odor of rotting flesh. Many investigators put cologne, orange juice, or Vicks VapoRub on their surgical masks to counteract the stench. Haueter put a sweet-smelling ointment called Tiger Balm in his moustache.

Supplee was haunted by the sharp smell of bleach. Each time the investigators left the crash site, they had to scrub their hands in a bleach solution to wash away any germs that might be present in the body parts. Supplee then washed his hands with soap to get rid of the bleach smell, but it would not go away. He smelled it every time he put a forkful of food in his mouth, every time he brushed his teeth. The Clorox seemed to be deep in his pores. He took shower after shower, but the smell lingered, triggering flashbacks of the carnage on the hill.

John Cox's biggest emotional challenge was picking through the cockpit. It was as if someone had destroyed the office where he worked every day. He did not know Emmett or Germano, but they were all part of the pilot brotherhood. Cox found skull fragments and parts of the pilots' brains on the autopilot panel, but he did not get emotional. He also found a finger with a ring still attached, but even that didn't bother him. The tragedy did not affect him until he pulled away a thicket of wiring and found Emmett's epaulets, the shoulder stripes that pilots wear to denote whether they are a captain or a first officer. The epaulets were still attached to Emmett's shirt, which was splattered with mud.

"One of ours," Cox said sadly. They had found Emmett, but it just as easily could have been anyone from ALPA. It could have been Cox himself.

He suggested that they take a ten-minute break, but the other investigators wanted to keep working. Cox said he desperately needed a break. He walked away, tears streaming down his cheeks.

The emotion erupted again during dinner at Mario's, an Italian restaurant that was a favorite of the pilots. Cox tried to convince his ALPA colleagues that he could work on both crash investigations—Flight 1016, the Charlotte accident that had occurred two months earlier, and the one in Hopewell. But the other union members were skeptical.

"You can't do this," one of the union officials said. "It's just too much."

"Let me find a way," Cox said.

They talked about what a challenging job it was. Cox found it rewarding—the hunt for clues, the idea that he could help solve the mystery of a crash, and the belief that he could actually make flying safer. But he was overwhelmed. He was working in a steamy rubber suit all day long and was getting only three or four hours of sleep each night. He started crying, right there in the corner booth at Mario's.

"All right," he said finally. "You're going to have to take me off of 1016."

The next day, he felt invigorated. The Mario's episode had cleansed him. "I hit the wall last night," he told his colleagues. "But I'm better now."

Many people in the Pittsburgh area regarded Beaver County as a Podunk kind of place. Ever since the steel mills shut down, it had been a bedroom community that emptied every morning as people drove to jobs in neighboring Allegheny County, which included Pittsburgh. Allegheny was bigger, richer, more sophisticated, and had the airport, the museums, the colleges, the Steelers, and the Pirates. Beaver County had the Hopewell High School Vikings.

No one expected the county coroner's office to be especially sophisticated at body identification procedures. The office was a throwback to the

1960s, with old furniture and a creepy opaque-glass door with CORONER in black letters—it looked like something from a Hitchcock movie.

In a typical year, the tiny office did only 100 autopsies. Two or three of those were unidentified bodies that were found in the woods, but otherwise the coroners knew the name of every dead body they saw. Suddenly they had 132 victims, with bodies torn apart worse than anyone had ever seen, and they had to identify them all.

The coroner's office sent six teams to the site to photograph and document the remains. Figuring that people sitting in different parts of the plane would be found in the same areas, the teams carefully recorded the location of each body part on the hill. They used a grid system of letters and numbers to indicate the placement, such as "IW32" or "KW920." Unfortunately, there proved to be no correlation between the location of the body parts and where passengers sat on the plane. Bodies had been blown in every direction.

The bagged body parts were stored in a refrigerated truck on the hill until they could be driven to the morgue at the 911th Airlift Wing, an Air Force Reserve unit at the Pittsburgh airport. A giant hangar normally used for big C-130 transport planes had been appropriated for the massive task of identifying the dead.

Wayne Tatalovich, the county coroner, accepted help from virtually anyone who offered—the Armed Forces Institute of Pathology, the FBI, and local hospitals and dental schools. As the Ziploc bags arrived, they were X-rayed to find any wreckage that had been mixed with the remains. An FAA bomb expert examined each body part for evidence of explosives. If there were bones in the bag, an anthropologist tried to determine if they were from a male or a female and attempted to estimate the person's age. Teeth were sent to dentists, who compared them against records submitted by the victims' families. Fingers went to an FBI team in the hangar that tried to take fingerprints.

The scale of the effort was staggering. There were 132 people on the plane, but 1,800 Ziploc bags. Workers on the coroner's teams made lots of mistakes. They wrote down the wrong letters for the grids where body parts were found. Their logs and photographs were inconsistent and incomplete. Their computers kept breaking down. Some were using Macintosh computers, others were using PCs, and the lists could not be transferred from one computer to the other. Volunteers who logged the findings into the database made repeated errors. In the space where they were supposed to list which personal effects were found with the remains, they often wrote, "Yes."

Tatalovich worked to keep the process as dignified as possible. Someone said prayers over the 911th's PA system. Once the body parts had been identified, chaplains escorted the remains to a hearse. Tatalovich made sure that the passengers' bloodstained money was replaced with new dollar bills from a local bank before wallets were returned to family members.

Each day a committee of pathologists met around a conference table to decide when they had enough information to identify someone. Most were identified through dental records or fingerprints, or both. If those methods didn't work, pathologists moved to more creative criteria—using skin color, the serial number of a hip replacement, wires in chest bones from open-heart surgery, or distinctive jewelry found on the body parts. The death certificates all said the same thing: "Accidental death due to severe blunt force trauma."

The process went slowly. A week after the crash, Joan's body still had not been identified.

To build an airtight case—if he ever came up with the cause—Haueter had to rule out every other possibility, no matter how far-fetched. So it was standard procedure to run drug and alcohol tests on the pilots. The tests nearly always were negative, but they had to be done to assure people that the pilots were not intoxicated. The tests were easy to perform as long as the pilots' bodies were intact.

In this crash, there were no bodies. Only parts. And Haueter had to be certain that the parts he was testing truly came from Emmett and Germano. Fingerprints and dental records had been sufficient to identify most passengers, but Haueter wanted to use DNA tests on the pilots to be 100 percent sure about the drug and alcohol results.

Genetic code known as DNA, which is unique in every human being, had been used to identify murderers and war victims. But DNA testing was relatively new in the early 1990s and had not been widely used for plane crashes. It seemed to be a perfect solution to Haueter's dilemma, however, because the tests could positively identify the pilots' remains.

The coroners from the Armed Forces Institute of Pathology who were helping with the investigation had body tissue that they believed came from the pilots—part of the upper left arm and back muscle for Germano and the back muscle for Emmett. But they had to find other blood or tissue from each pilot or a relative so they could match the sample with one that was known to have his DNA. Emmett had no children, but his mother was still alive and her DNA would be similar to his. She agreed to give a blood sample to a local doctor. When the experts compared her blood with the DNA from the back muscle, it matched well enough that the experts were sure the muscle had come from Emmett.

The NTSB ran into difficulty getting a match for Germano, however. His parents were no longer alive, and his wife did not want to provide blood samples from their children. Then someone in the investigation remembered a foot that had been found in the cockpit area. The FBI could match it with a footprint taken of Germano when he was in the air force. Unfortunately, by the time the pathologists realized it could be used for DNA, the foot had been placed in a casket to be sent to Germano's family. Haueter quickly called Tatalovich.

"I need a piece of the foot," he said.

They just closed the casket, Tatalovich told him.

"I need a piece of that foot. Open it and clip off whatever the AFIP guys need and take it to them." Haueter could not believe his own words. He was making decisions about a dead man's foot.

Tatalovich was concerned that he wouldn't have much to send back in the casket to the family.

"Don't take out the whole foot, just take off a little chunk," Haueter said. Tatalovich agreed.

When they compared Germano's foot with the footprint, it matched. Then they compared the DNA from the foot with the muscle. It matched.

The muscle specimens for both pilots were then sent to the FAA's Civil Aeromedical Institute in Oklahoma City for analysis; no drugs were found. The tests disclosed a small amount of alcohol in both samples, but that would have been caused by the natural chemical changes in the muscle since the crash. Haueter was now sure that the pilots had not been intoxicated at the time of the crash.

8

THE PSYCHIC AND THE

DRUG DEALER

Paul Olson, the man who sat in Seat 17F, was a convicted drug dealer. He had started selling marijuana as a teenager and then switched to cocaine when marijuana sales declined. He was a smart businessman in the thriving South Florida market, adjusting his product line when demand changed. He was a wholesaler. Suppliers dropped bundles of marijuana or cocaine a few miles offshore in the Atlantic Ocean, and Olson fished them out and sold them to other dealers. He was said to have been a millionaire by age twenty-one.

But the lavish life came to an end. Olson got caught, was convicted, and had to serve five years in federal prison. When he got out, he said he had turned over a new leaf and had become a law-abiding citizen. He took a job as a driver with a seafood company. His only link to the past was a requirement that he testify against other drug dealers.

When federal prosecutors called him in the summer about testifying against the alleged leader of a notorious Chicago drug ring, Olson told his fiancée that he was nervous about it. "I thought that part of my life was behind me," he said. Yet he seemed relaxed when he arrived at the Chicago offices of the U.S. attorney on September 8. He wore a T-shirt and a baseball cap. He had a good tan. He and the prosecutors talked for a few hours, discussing his possible testimony, but they decided he wouldn't be much help. The meeting ended early, and Olson, who had been scheduled on another flight, was rebooked on Flight 427.

After the crash, an assistant U.S. attorney called USAir and told them about Olson. Haueter got the news a day or two later but was assured by the FBI that Olson would not have been a target for a hit man. He was small potatoes. He was never part of the Witness Protection Program, in which the federal government finds new homes and identities for key witnesses. It was simply that the terms of his drug conviction required him to testify in other cases.

But Olson made for a great conspiracy theory. *Maybe someone had blown up the plane to silence him!* As the week wore on without the NTSB finding an apparent cause, the conspiracy theories picked up steam. That's how a small but vocal segment of people deals with uncertainty. In the absence of concrete answers, they blame the forces of evil—the Mob, the Trilateral Commission, the CIA. And just for good measure, they say the federal government covered up the whole thing.

The press, which was getting less and less from the nightly NTSB briefings, pounced on the story. FLIGHT 427 VICTIM LINKED TO DRUG CASE, read the headline in the *Chicago Tribune*. The Gannett News Service called him a "mystery federal witness" and quoted attorney F. Lee Bailey as saying, "I believe a bomb caused it, particularly after I found out what went on in the cockpit. In my view, in the totality of what we know so far, the scenario is consistent with a bomb. I think two bombs might have gone off in sequence." The bomb theory was fueled by a *New York Times* story the same day that quoted an unnamed "aviation official in Washington" who described a mysterious "whoomp, whoomp" sound followed by a surprised grunt from one of the pilots and a voice in the cockpit asking, "Jeez, what was that?" (That mistake in the *Times* story—Germano actually asked, "What the hell is this?"—was typical of press coverage in the early days after a crash. Reporters were so eager to get scoops that they relied on anonymous sources—in this case someone with a secondhand account of the tape who got important facts wrong.)

Reporters kept badgering Haueter with questions about Olson, even though the FBI had said publicly that there was no evidence that anyone had tried to kill him. Finally Haueter got frustrated and called the FBI again. Was there more to the Olson story than he had been told?

No, the agent said. "He's nobody."

The early clues from the crash argued against a bomb. The plane's wreckage, including the four thousand copies of *BusinessWeek*, was concentrated in a tight area on the hill. If a bomb had gone off in midair, *BusinessWeek*s would have rained from the sky for miles. They would still be finding copies in Cleveland. But Haueter couldn't be sure. He needed to rule out every possibility, no matter how silly it was. So he asked Ed Kittel, the FAA's bomb expert, to do a thorough examination of the wreckage.

Kittel was one of the world's premier experts on airplane explosions. He spent twenty years as a bomb disposal officer for the navy, including time that he vaguely refers to as "working with the intelligence community." He had intense blue eyes, a neatly trimmed beard, and short hair that was still cut to navy standards. On his desk in the FAA's Washington headquarters was a picture of his two-year-old son with an antique detonator. His computer's screen saver read, "There's no problem that cannot be solved by the proper amount of high explosives."

Kittel knew that bombs left a calling card. If one had exploded inside the USAir plane, hot gas from the bomb would have spread quickly, deforming everything it touched. The wreckage would be pitted with tiny holes that resembled craters on the moon. It would also have black streaks that would radiate from the origin of the blast and would not wipe away with a finger. By comparison, a fire after the plane hit the ground would be cooler and slower. The metal would not get pitted. The soot would wipe away with a finger. The problem for Kittel was that the 737 was in tens of thousands of pieces. He had to check every one.

So he positioned himself on the hill at the place where wreckage was sprayed with Clorox and checked each piece of wreckage like a supermarket cashier in a moon suit. For days, he kept at it, inspecting virtually every piece. He held them, turned them over, and searched for the telltale pockmarks. He found no sign of a bomb.

The wreckage was then trucked to a USAir hangar at the Pittsburgh airport and spread out on the floor in the rough shape of the plane. The idea was to look for burn patterns and broken metal that might indicate an in-flight fire or bomb. But once again, Kittel saw no evidence of either. The burn marks were randomly scattered around the plane, which meant the pieces had not caught fire until the plane struck the ground and the parts mixed together.

At the makeshift morgue at the Air Force Reserve hangar, Kittel's partner Cal Walbert examined body parts for evidence. If a bomb had gone off in the cargo compartment, pieces of it would have been captured in the passengers' legs and buttocks. The coroners X-rayed every Ziploc bag, removed any foreign matter, and placed it in recycling bins. Walbert sifted through twenty bins and found fragments from seats, plastic trays, and overhead bins, but nothing to indicate a bomb.

Kittel told Haueter that he was positive: A bomb had not brought down Flight 427. But that night at the Holiday Inn, a reporter ambushed Haueter as he came out of the press conference.

"I know this is a bomb," the reporter said.

Haueter grabbed Kittel, introduced him to the reporter, and said, "Ed, is there any doubt in your mind that this was not a bomb?"

"None," Kittel said. "We have no indication of a bomb of any kind."

The crackpots had started calling on September 9, blaming the crash on the devil, Russian death rays, and the Prince of Darkness. The calls were dutifully forwarded to the NTSB witness group. The group had two purposes: to field calls from people with theories about the crash and to interview anyone who had seen Ship 513 fall out of the sky.

Witnesses to plane crashes were notoriously unreliable. They often confused the order of events and embellished their stories. But interviewing them was a standard part of every investigation, and occasionally someone actually saw something important. The 427 group got names of witnesses from newspapers and TV news reports and then tracked them down, taking a plastic model of a USAir plane so people could demonstrate what they had seen. Most said the plane had rolled left and then plunged nose down toward the hill, as the flight data recorder indicated. But they could not agree on whether it made an unusual noise. Some heard a growling sound. A kindergartner thought it sounded "funny." Others heard nothing out of the ordinary. A USAir utility worker who happened to be on the soccer field behind Green Garden Plaza was sure that he saw smoke coming from the front of the right wing. Others saw no smoke. One man saw a mist and believed the plane might be dumping fuel. A food service employee said one of the engines looked like it was cocked to the side. Other witnesses said the engine was fine.

The witness group got lots of calls from retired airline pilots and well-meaning frequent fliers who said the crash reminded them of problems they'd experienced on other flights—some that were years earlier. One man said an electromagnetic field from power plants might have harmed the 737's "fly-by-wire" system (he apparently did not know the 737 was not a fly-by-wire plane). A retired pilot from Stuart, Florida, faxed an elaborate scenario suggesting that a flap failure surprised the pilots. He ended his scenario with "Precious little time and mind frozen with fear."

Another caller said he had a dream that a flight attendant was standing in the cockpit and fell on the captain when the plane banked, causing the captain to lose control. The man said only pilots should be allowed in the cockpit. If that wasn't feasible, others in the cockpit should wear a restraining belt to keep them from falling into the pilots. The NTSB witness group kept a log of each interview, noting whether a follow-up was recommended. In the follow-up column for the man with the cockpit theory, one of the witness team members wrote, "Possible psychiatric help."

Haueter got a postcard that read, "Give me $50,000 and I'll ask the Prince of Darkness for no more plane crashes." Haueter joked that it was a risky investment. The guy wasn't promising the crashes would stop. All he said was that he would *ask*.

Witness group member Kimberly Petrone, a USAir flight attendant, took a call from a man who said he was a psychic and had a dream about the accident. He had envisioned a clear blue sky and then the word "hydraulic" written in the clouds.

Petrone couldn't take the guy seriously. "What's the lottery number going to be tomorrow?" she asked.

"I'm not that kind of psychic," he said. "I just sense things."

This is the paradox of aviation history: Birds inspired us to build the first airplanes a century ago, but an errant flock could bring a 737 crashing to the ground and give us a humble reminder that humans were not meant to fly. Birds and planes had an uneasy coexistence from the start. Calbraith Rodgers, the first man to fly across the United States, was also the first to be killed by a bird collision. His Wright brothers plane struck a gull in 1912 and crashed in the surf at Long Beach, California. Since then, bird-plane collisions have killed more than one hundred people in the United States and countless others around the world. The 1960 crash of a Lockheed Electra in Boston, which killed sixty-two people, was caused by a flock of starlings and gulls, and a 737 crash in Ethiopia that killed thirty-five people was blamed on speckled pigeons.

A few birds of modest size could do serious damage to a $40 million plane, breaking through the nose or the windshield or crippling the engines. When a four-pound bird hit a plane going 260 knots the bird had the force of fourteen tons.

The FAA gets two thousand "bird strike" reports in a typical year, but an estimated eight thousand strikes never get reported. Gulls account for one-third of the collisions, followed by ducks/geese/swans, blackbirds, doves, and raptors. (The FAA also keeps track of ground collisions with other wildlife. In a typical year, deer are struck by planes forty-three times. Other collisions involve coyotes, dogs, skunks, muskrats, and possums. Reptiles are a smaller risk, although planes struck two alligators from 1992 to 1996.)

At the airports with the worst bird problems, employees drive around in trucks playing tapes of screeching birds over loudspeakers. The tapes sound like a flock of starlings being tortured. If that doesn't scare the birds away, the workers pull out shotguns and fire blanks into the air. With stubborn birds, the employees occasionally shoot to kill. The world experts on bird-plane problems work for an air force unit called the Bird Aircraft Strike Hazard (BASH) Team. The BASH Team visits air force bases to warn pilots about the problem and make sure airfields are not attractive to birds. They tell fighter pilots that a single turkey vulture can be as dangerous to an F-16 as a round of artillery.

The FAA sets standards to make civilian planes bird-resistant, saying they must withstand the impact of birds as heavy as eight pounds. Tests to certify the engines look like a bizarre stunt from a David Letterman show. The engineers put dead birds in a cannon and shoot them into the spinning fan of a jet engine.

At the crash site, theories about birds emerged in the first few days. A bird collision might account for the mysterious thump on the cockpit tape, and it might explain why the plane suddenly rolled to the left. Maybe a big goose broke through the skin of the plane and jammed the rudder system. Or maybe it got lodged in the slats, the movable panels on the front of the wings.

John Cox was skeptical. In his twenty-year career as a pilot he had collided with several birds, and he knew they could be dangerous. But he doubted that birds had caused this crash. He kept saying, "Have to be a hell of a bird to bring down a 737."

No feathers were found in the wreckage, and air traffic controllers said they saw no birds on radar at the time of the crash. But five witnesses said they had seen birds in the area the previous evening or the morning of the accident. A retired army colonel called to say that he had seen "thousands" flying over Hopewell three nights earlier. The operations group checked with a bird expert at the University of Illinois, who said it was unlikely but possible for birds to be at 6,000 feet.

At the hangar where the wreckage was being assembled, investigators walked up and down the makeshift aisles, looking for anything birdlike. They paid special attention to the wings and the nose, where birds were most likely to hit. But the only evidence they found was a pair of feet from a goose that someone had left in one of the engines as a prank—a stunt that infuriated Haueter. Dozens of people studied the wreckage, but no one saw any sign of a bird.

That didn't satisfy the team from Boeing, which was especially interested in the bird theory. (Haueter noticed that Boeing was always keenly interested in theories that shifted blame away from its airplane.) Rick Howes, Boeing's coordinator, reminded Haueter of the destructive power of buzzards and how they had caused several crashes. Howes said he was interested in the left wing slats, where a broken hinge might have been caused by a bird. He thought the NTSB should do a full-scale reconstruction of the leading edges of the wings and the area just behind the nose, known as the forward pressure bulkhead. Once those sections were reassembled, the investigators could examine them with a special black light that would make bird residue glow.

Haueter was pretty sure birds weren't a factor, but he wanted to be positive. He agreed to do the reconstruction. "Let's bring the theory up now and bury it," he said. "I don't want to have it haunting me a year from now."

Greg Phillips was nearly killed investigating a 737 crash. A flight controls specialist with the NTSB, he had been sent with Haueter to Panama in 1992 to figure out why a Copa Airlines 737 had suddenly twisted out of the sky at 25,000 feet and crashed. As he searched the jungle for wreckage, Phillips felt a sting on his arm and slapped away what he thought was a caterpillar. Back in his hotel room, he began to feel ill. His stomach ached, his fever rose, he shook with chills. He felt like he was falling down a long, dark hole. Lying there naked, he thought he was going to die. He crawled out of bed and put on some pants. If he was going to die, he was going to go wearing pants. He recovered sufficiently to take the first flight home to Washington, and blood tests showed he had been bitten by a black scorpion, which was often fatal.

Phillips got interested in accident investigation as an engineering student at the University of Evansville in Indiana. In 1977 the school's basketball team had been killed in the crash of a DC-3 because of a problem with the plane's rudder. Phillips read everything he could about the crash and decided he wanted to be an investigator. After he graduated, he designed airplane components for Cessna in Wichita, Kansas, and for Northrop in Los Angeles until he saw an ad for an NTSB job in *Aviation Week and Space Technology* in late 1987.

He had frizzy brown hair, button eyes, and a dimple on his chin. He always sounded happy, even when he was buried in work and about to be sent on a fifteen-hour flight to yet another crash. Like his friend Haueter, Phillips loved to put things together. He was building an RV-8, a lightning-fast experimental plane that could be used for aerobatics. He drove a Porsche 944 and a Harley-Davidson motorcycle and brewed his own beer. Phillips had the perfect personality for the safety board's culture of caution. He was thorough and always skeptical. When other investigators were convinced about some kind of malfunction, Phillips was often the last holdout, constantly looking for more evidence.

Many investigators came across as dispassionate. They were like homicide detectives. They didn't get wrapped up in the tragedies they saw because every day brought a new one. But Phillips wasn't afraid to discuss his feelings about what he saw in the field. He said his job reminded him how fragile life is. "There's no guarantee for the next day," he said. "You can't put off expressing emotions to people you love." After an accident he often spent extra time with his wife, Debbie, and his two teenage boys, savoring each moment with them.

Phillips was blunt about the limits of technology. "You can't have a perfect airplane," he said. "I don't think perfection exists." The systems on modern jetliners were so complex that no human being could possibly account for the myriad ways in which things might fail. Yet he never worried about his own safety when he flew. He felt that the manufacturers built solid airplanes, that pilots were well trained, and that the government did a good job of regulating the airlines and the manufacturers. "The system works," he said.

When he arrived on the hill in his rubber suit, the largest piece of wreckage he saw was the tail. On the wreckage map, the tail was piece No. 1, right at the center, like a big **X** on a treasure map. The rest of the plane had been squashed and shredded by the impact, but the tail was in surprisingly good shape. That was good news to Phillips because two of the primary flight controls were on the tail: the rudder (the movable vertical panel that made the plane's nose go right or left) and the elevators (the horizontal panels that made the nose go up or down). Fire had devoured an eleven-by-four-foot section of the vertical stabilizer, which was the big fin area in front of the rudder, and there was a chunk missing from the top portion of the rudder panel itself. But inside the tail, the hydraulic devices that moved the rudder were in good condition.

Phillips's approach on the hill followed the classic NTSB method. He had his team examine each component from the flight control system, even if the component was unrelated to the rudder. It was far too early to jump to conclusions. They did not want to zero in on one theory too fast, the way the investigators of the Knute Rockne crash did, only to find later that they had neglected something important.

His team found the other mangled flight controls—the ailerons (the wing panels used for banking and turning), and the flaps and slats (the devices on the front and rear of the wings that provided extra lift during takeoffs and landings). For each one, they also found the actuators—the hydraulic or mechanical devices that moved the panels. Phillips's team carefully measured and photographed everything, to record whether the panels were up or down, right or left at impact. By measuring where things broke or where they bent, they were able to put together a snapshot of what the plane was doing when it struck the road.

The group spent a lot of time examining the device that moved the rudder, which was called the power control unit. Amazingly, it had survived the crash with virtually no damage.

"This thing is in pretty good shape," Cox said when he saw it.

Phillips was relieved. He knew that the rudder unit would be the subject of many tests. One of his biggest regrets from the Colorado Springs crash was that the rudder system had been badly damaged, making it difficult to test some of the theories. Phillips's first goal with the USAir wreckage was preservation. He needed to make sure the evidence wasn't altered when his team moved it from the hill to the hangar and then shipped it to a lab for a detailed inspection.

A yellow construction crane was brought in to lift the big tail section onto a flatbed truck. Workers tied nylon straps around it and stood back as the diesel engine strained to pick it up. When Cox heard the engine groaning, he worried for a moment that the crane might drop the huge piece. But the tail lifted slowly into the air, and the workers used ropes to direct it onto the truck. Once it was secured, the truck drove to the decontamination station so it could be sprayed with the Clorox solution. The process reminded Cox of his planes' being de-iced. The wreckage was covered with a tarp to hide it from the snooping eyes of the media, and then it was driven to the hangar.

Phillips and his group decided to send the key pieces to two labs on the West Coast. Most devices from the rudder and aileron systems would go to the Boeing Equipment Quality Analysis lab outside Seattle, which was regarded as one of the best places in the world for forensic analysis of a plane crash. The rudder power unit would go to a Parker Hannifin plant in Irvine, California, where the units were manufactured.

The rudder unit had been carefully cut from the tail of Ship 513. Its hydraulic lines were capped to keep the fluid inside. It was packed in a padded trunk and sealed with several layers of shipping tape. Phillips and other members of the group signed their names on the tape so it would act like police evidence tape. When Phillips got the box on the West Coast, he could check the tape to be sure that no one had opened the trunk and tampered with the evidence.

9

PIPSQUEAK

Hidden in the tail of every 737 was a hydraulic device that acted as the muscle to move the plane's big rudder. It was about the size and shape of an upright vacuum cleaner and was known as the power control unit, or PCU.

Inside the PCU was an ingenious valve. It was shaped like a soda can and was strong enough to move the rudder when the plane was going 500 miles per hour. The gadget had an impossibly dull name—the dual concentric servo valve—but engineers used an amusing hand gesture to show how it worked. They curled the fingers of one hand to create a hole and then stuck their index finger in and out, as if they were demonstrating sex.

That was how it worked. When a pilot pushed on a rudder pedal, he moved a tube the size of a pencil in and out of the soda can. Holes in the tube allowed hydraulic fluid to squirt against a piston, which pushed the rudder to the right or left.

The 737 rudder valve was unique because it had *two* tubes, one inside the other. Both moved back and forth. That design was revolutionary in the arcane world of hydraulics. It allowed Boeing to use one power control unit instead of two, which helped it fit in the tight confines of the tail and trimmed at least fifty pounds off the final weight, saving airlines thousands of dollars in fuel over the life of a plane.

The valve-within-a-valve also had an important safety feature. If one tube jammed, the other one could still move and neutralize the rudder.

At least, that was how it was supposed to work.

Investigators had suspected a problem with the rudder valve in the Colorado Springs crash, but the evidence had been too sketchy. The valve was damaged and difficult to test. This one, however, was in perfect condition. So Phillips's trip to the West Coast had the potential to wrap up the investigation quickly and neatly. All he needed was to find a tiny flaw, and the mystery would be solved.

But as he flew to Seattle, he wasn't fantasizing about solving the case. He was being his usual cautious self, fretting about the test plan for all of the flight controls—not just the rudder—to make sure he didn't forget anything.

"We can't just stay focused on the rudder," he told his group when they gathered in a lab at Boeing's big plant in Renton, Washington. There was mounting evidence pointing to the rudder, but Phillips did not want to exclude other possibilities and then discover he had missed something crucial. He said he would be open to any idea, any test. "No question is too trivial."

The lab, in a giant gray building beside Interstate 405, about ten miles southeast of Seattle, resembled a high school science classroom, with drab yellow walls and lots of counter space for tests. The technicians who worked there were the unsung heroes at Boeing. They were a notch lower on the food chain than the engineers who got most of the glory, but they were just as important. They tested hundreds of airplane parts every year to make sure they were safe and durable. They could figure out why a window cracked, why an actuator leaked, how to improve a toilet seat.

They had lots of gadgets to help them examine broken parts—microscopes that let them see tiny imperfections in plastic and borescopes that could peek inside a device. The technicians were like a band of renegade nerds. For kicks, they once used the borescope to check out a guy's ingrown nose hair.

Phillips's group planned to test anything that could have played a role in the crash—the main rudder power unit and its backup, the standby PCU; the rudder trim actuator, which adjusted the rudder constantly to keep the plane flying straight; and the hydraulic devices that moved the ailerons. The wreckage arrived in big wooden crates, all sealed with tape, just as the rudder crate had been, to make sure no one tampered with the evidence. Everything reeked of Clorox.

Phillips, Cox, and the other team members met in a conference room to decide how each piece would be tested. Phillips insisted on a consensus for every step. They would not proceed until all team members—from the NTSB, the FAA, Boeing, ALPA, and the machinists union—agreed.

They methodically inspected and tested each item. Some parts were taken to another Boeing building and X-rayed to search for internal cracks. The team was looking for anything out of the ordinary—scrapes, leaks, dents, electrical fluctuations—that might explain what had gone wrong on Ship 513.

One theory about the crash was the possibility of "runaway trim" of the rudder. Normally, pilots could turn a dial to trim the rudder to adjust for crosswinds and keep the plane flying straight. But an electrical malfunction might have made the trim "run away," sending the rudder swinging to one side. The trim actuator, about the size of a box of animal crackers, was badly dented from the crash. But one of the Boeing technicians figured out a way to hook a testing device to it. When he measured the amount of electricity that went in and out of the device, the electrical readings showed the trim was centered. Runaway trim had not caused the crash.

They examined dozens of other pieces the same way, taking photographs and carefully documenting the condition of each piece. On the hydraulic pressure gauge, a tiny indentation known as a "witness mark" provided valuable information. When Cox found the gauge at the crash site, it had one needle pointing to 3,000 pounds per square inch (psi), which showed that the "B" hydraulic system was operating normally when the plane crashed. Unfortunately, the "A" needle had broken off. But when the technicians put the gauge under a microscope, they could see a distinct mark indicating that the A needle had also been pointing to 3,000 psi. That meant both systems were working properly and they could rule out a hydraulic failure.

After scores of tests, they came up empty. Everything they tested was working fine when the plane struck the road. It was time to go to Irvine, California, and see what they could learn from the soda can valve.

At the time of the Hopewell crash, Boeing was number one in a dwindling group of airplane makers. Lockheed and Fokker were no longer building airliners, and McDonnell Douglas was struggling. Boeing's primary competition was Airbus Industrie, a consortium of European companies based in Toulouse, France.

In Seattle, Boeing was the city's largest private employer and a mighty economic force. You couldn't drive far on a major highway without seeing the company's familiar italic logo. The major plants were along East Marginal Way on the edge of Boeing Field; in Everett, a town north of Seattle where the wide-body jets were assembled; and south of Seattle in Renton, where the 737s and 757s were made.

Boeing sometimes lagged behind its competitors in developing new planes—a practice that earned the company the nickname the Lazy B—but it seemed to gamble at the right times. In that regard, the 737 was typical. When Boeing officials were debating whether to build the plane in the early 1960s, archrival Douglas was already several years ahead with the similar-sized DC-9. Boeing president William Allen was dubious about building the 737 because his chief competitor was so far ahead. He had struggled through the high costs and development problems of the three-engine 727 and was not sure the company could afford another painful birth so soon.

The 737 was designed for short-haul flights of five hundred to a thousand miles, serving airports that otherwise might not get jet service. Boeing planned to offer customers an optional gravel kit, so the plane could land on an unpaved runway. Early designs had the engines on the tail, as on the 727, but they were later moved to the wing because placing them there would allow seats for six more passengers.

By 1964 many Boeing executives felt that the 737 project was dead. But a handful of believers led by chief engineer Jack Steiner pushed to keep the project alive. They said costs could be reduced because the plane would be heavily based on the 727. The new plane would have many of the same parts as its sibling.

Designers tried to simplify every aspect of the 737 to save money and increase reliability. By combining two parts into one—such as the rudder's unique valve-within-a-valve—the designers would have fewer things that could break. That was crucial if they were to achieve the ambitious goal of 99 percent reliability. They also simplified the cockpit and pilot workload so the plane could have a two-person crew instead of the costly three that were needed for the 727.

Steiner, an intense workaholic who wore Buddy Holly glasses, decided to sneak behind his boss's back to convince the board of directors to support the 737 launch. It was easy for Steiner to reach board member Ned Skinner, because he lived nearby; Steiner simply walked over to Skinner's house and gave him the sales pitch. Then Steiner tracked down three other board members and gave them the same message. The lobbying campaign worked. The board decided in 1965 to build the plane.

When the first 737 came off the assembly line in Renton in January 1967, it was christened by seventeen stewardesses with seventeen bottles of champagne, representing the number of airlines that had ordered the plane. When Allen spoke at the dedication, he noted that the small plane had been overshadowed by bigger jets being developed—the 747 and the supersonic transport. The 737 might be small, he said, but it would fill an important niche for shorter flights. "We expect the 737 to serve long and well on the air routes of the world."

It took nearly twenty years for the plane to make a profit, but the 737 is now the best-selling jetliner in the history of commercial aviation. Boeing boasts that its 99.2 percent reliability is the best in the world. There are more than 3,800 of the planes in use, with 700 in the air at any given moment. The 737 is so successful that Boeing now makes the "Next Generation" models, with longer ranges, more-efficient wings, quieter engines, and more-modern cockpits.

The plane that crashed in Hopewell was a 737-300, a model that some pilots called the Quichewagon. The 300-series was the first to have a flight-management computer, which enabled pilots to enter complicated routes into the computer so the plane could virtually fly itself. The computer helped them find the most fuel-efficient altitudes, saving up to 7 percent in operating costs. Pilots who relied on the high-tech device were said to be less macho. They were the quiche eaters.

The 737 was dull but efficient, with all the sex appeal of a four-door sedan. Other planes had mean-sounding nicknames like Mad Dog and Mega-Dog, but the 737's monikers sounded downright wimpy: Fat Albert, Guppy, and Fat Little Ugly Fellow (FLUF). Boeing was famous around the world for its gigantic 747, the premier long-haul airplane. When a 747 took off from a runway, people stopped and watched the 400-ton machine climb into the sky. But the smaller 737 rarely stopped traffic, since so many of them were making so many flights. Watching a 737 take off was about as exciting as watching a minivan pull out of a garage.

The lackluster image of the 737 was largely a function of its plain-vanilla use. The big 747s flew twelve-hour missions on long, glamorous routes such as New York to Paris or Los Angeles to Honolulu. The pipsqueak 737s flew short hops like Baltimore to Buffalo or Detroit to Little Rock. The 737 was a relatively basic machine. New planes were "fly-by-wire," with computers sending electrical signals to move the flight controls. But the 737 still relied on old-fashioned cables that ran beneath the cabin floor.

The plane had an excellent safety record, but so did virtually every other modern plane. The 737-300, -400, and -500 planes had nine "hull losses" (the industry euphemism for a crash), for an accident rate of 0.43 for every million flights. The 737's rivals, the DC-9 and the Airbus A320, had higher rates, but the differences were insignificant. Crashes were rarely caused by the same mechanical problem. About 70 percent of crashes were blamed on pilots, and only 10 percent on malfunctions.

Yet two or more crashes caused by the same problem in a plane could destroy the plane's image and make airlines stop buying it. A design flaw that led to crashes of the de Havilland Comet in the 1950s nearly doomed the plane. The same thing happened to the Lockheed Electra in the early 1960s after a series of accidents were blamed on engine vibration. The DC-10's reputation deteriorated so much after two spectacular crashes in the 1970s that airlines began canceling orders.

So Boeing had a lot at stake in the Flight 427 investigation. There now had been two mysterious 737 crashes with similar circumstances. Boeing had to reassure its jittery customers that its plane did not have a fatal flaw.

Some airlines were so nervous about the crash that they instantly reacted to every theory that came up. When the NTSB raised the possibility that one of the plane's thrust reversers had suddenly deployed in flight, a few carriers immediately called and said they wanted to disconnect them. Boeing officials sent telex messages to the airlines nearly every day for the first two weeks, telling them in one message, "We would like to stress that, contrary to some media reports, at this point the NTSB has not concluded there was a partial in-flight thrust reverser deployment. Boeing has no recommended operator action at this time. If the investigation shows any specific actions are recommended or required, operators will be notified."

Boeing had a special team to deal with crashes—Air Safety Investigations, which was a select group of engineers run by a shrewd manager named John Purvis. The team's mission was to help the NTSB and determine whether Boeing needed to fix any problems with the plane. For Flight 427, Purvis's office was overseeing an unprecedented effort involving more than a hundred Boeing employees. Engineers who designed flight controls were studying the 737 rudder system for possible flaws. Aerodynamic experts were studying the numbers from the flight data recorder, trying to estimate when and how far the rudder had moved. The company's structures experts were helping the NTSB make sense of the wreckage.

The crash was so important to Boeing that the chief engineer for the 737, Jean McGrew, had been appointed to work with Purvis and oversee the company's response. McGrew said his mission for Flight 427 was simple: to help the safety board find the cause of the crash, and—if the 737 was to blame—fix it. It bothered him that some people thought Boeing might be trying to cover up flaws in the plane to save money. McGrew said he was open to any theory, even if it meant Boeing might be at fault.

From his boyhood in Missoula, Montana, McGrew seemed destined to be an aviation engineer. He delivered the *Missoulian* newspaper and used his profits to buy model airplanes. He loved the challenge of designing and building something that could actually fly. He started his career with the Douglas Aircraft Company as a young expert on flutter, the potentially catastrophic condition that can make an aircraft shake so violently it breaks apart. He liked flutter because it was complicated. Some engineers preferred a single, more simple discipline, but McGrew chose flutter because it involved virtually every aspect of aerospace engineering—structural mechanics, stress analysis, design, and aerodynamics. The more complicated, the better.

He was an early computer expert who introduced Douglas engineers to the power of the PC. He rented an Apple II and holed up in a Manhattan hotel room, crunching numbers to prove that their new MD-80 jet could safely land on LaGuardia Airport's runway pier. He also was one of the first guys at the company to have a computerized Rolodex.

In 1989 he retired from McDonnell Douglas and jumped to Boeing, which was the aerospace equivalent of switching from Pepsi to Coke. He became the chief engineer for the 737, which meant he was the plane's godfather, responsible for any design changes.

He was so smart that his mind seemed to leap ahead during meetings, solving problems that weren't even on the agenda. His mind was like a mainframe. He studied a problem, processed the data, and reached a conclusion. One of his kids once gave him a gold Slinky on a plaque that said, RELAX, a reminder that he often got wrapped up in his job, but he rarely had time to play with it.

McGrew had a wry sense of humor and a love of tennis, boating, and spicy food. He solved problems with a cool, analytical approach and calculated everything to close tolerances. He knew precisely how long it took to get to the airport and always cut it close so he wouldn't be sitting at the gate wasting time. He fit well into the Boeing engineer-dominated culture. He looked the part—a thin fifty-six-year-old man with wire-frame glasses and a pocket protector that he kept hidden inside his shirt pocket, as if he were slightly embarrassed about it. He loved computers, but he was not the kind of guy who would surf the Web or play games. To McGrew, computers were tools, not toys.

His decision about wearing a pocket protector was calculated with the same logic he used for everything else. He wanted pens and pencils within reach, but he didn't want stains on his dress shirts. Never mind if the pocket protector looked nerdy. Data in, data out. A decision was made, his pockets were protected.

Boeing and USAir kept elaborate records on everything that was done to every plane, and Ship 513 was no exception. There was paperwork virtually every time a mechanic touched it, even for the routine transit check the night before the accident. The papers said the mechanics at the Windsor Locks, Connecticut, airport had examined the engines, checked the brakes and tires, switched on the exterior lights, and checked the galleys and lavatories.

The NTSB examined the maintenance records and found that 513 had suffered the usual ailments and hiccups of a seven-year-old plane—a blown tire, a fuel line problem, tiny cracks in a floor beam, a few spots of corrosion—but nothing extraordinary. Investigators scoured maintenance records and found a few "carryover" items, minor problems that did not need to be fixed right away. The engine thrust reversers were showing some wear. And the cabin floor was soft and spongy at Row 5 and was fixed with a temporary patch.

A year before the crash, the plane had undergone the rigorous USAir "Q check," a three-week exam in which mechanics inspected every beam and panel. The plane had gotten the requisite minor checkups since then, and it complied with the latest safety requirements from the FAA. Ship 513 seemed to be in great shape.

Cox checked his records and found he had actually flown it several times, but not in the last several months. Each 737 had its own quirks. They felt different to pilots because of their age, their individual equipment, or because they occasionally got banged by a passenger walkway. Some were cranky and had a rougher ride than others. But 513 was known for a smooth and steady ride. When Cox spoke with a colleague who had flown it recently, the pilot said he was sorry 513 was gone.

"It was a good one," he said.

When the systems group arrived at the Parker Hannifin plant in Irvine, Cox, the ALPA representative in the group, could feel tension in the air. To the Parker employees, this was like a visit from an IRS auditor. The NTSB group was there to see if the rudder power control unit—Parker's bread and butter—had caused a crash that killed 132 people. Employees were jittery.

With the help of Parker technicians, the NTSB team pulled the silver-colored unit out of the crate and methodically took it apart, piece by piece, examining each one under a bright light. They checked one of the filters and found nothing out of the ordinary. A group member acted as the scribe to document the tests. He wrote, "No large particles."

They checked the O-ring seals on the filters. The scribe wrote, "No deficiencies." They opened one end of the power unit and peered inside with a borescope. "No evidence of impact marks or abnormal wear marks." They opened another cover plate and peered inside at the fluid.

It twinkled.

Cox could see tiny pieces of metal in the fluid, glittering in the lights. "Is this normal?" he asked another team member. The guy shrugged.

The question circulated through the group. Was it supposed to twinkle? Wasn't hydraulic fluid supposed to be cleaner? Finally, someone from Parker piped up. The area was downstream of the metal parts, and it wasn't unusual to get tiny particles there. Still, no one knew how much twinkle was too much. They looked inside with a microscope and videotaped the tiny particles. Then they drained the fluid with a syringe and sent it to be tested.

The next day, they hooked the power control unit to a test bench that looked like a refrigerator turned on its side with dials and switches on the back. The bench acted like a hydraulic pump on an airplane, providing pressure to the unit. Some members of Phillips's team believed the test could bring their first breakthrough. The valve-within-a-valve might be jammed, which would make its piston go hard to one side once hydraulic pressure was turned up. That would explain the crash.

Expectations for the test were so high that Phillips had invited Carl Vogt, the chairman of the safety board, to watch. Vogt and everyone in the systems group stood around, intently watching the piston. A technician drained the last of the fluid from the unit and collected it in a clean container so they could send it to Monsanto, the manufacturer. They installed a special cover on the unit so they could watch how the valve and the levers inside reacted. They would be able to see if it went haywire.

A technician flipped a switch and started the purple hydraulic fluid flowing at 360 pounds per square inch, about one-tenth the normal pressure. The bench vibrated with a steady hum of the pump. No one spoke. Everyone was listening for a hiss or gurgle or belch or some other strange sound that the valve might have made before it threw 132 people to a horrible death. It seemed as though everyone in the room was holding his breath, waiting for the big moment. Everyone craned to see the valve, watching for a jam or a sudden movement.

Nothing.

The piston just sat there. The valve didn't budge. They might just as well be staring at a rock garden.

A technician moved a lever to simulate the pilots' stepping on the rudder pedals, and the slides inside the valve went back and forth smoothly. The scribe wrote, "Servo valve acted normally." They turned up the pressure to a full 3,000 pounds per square inch, the normal pressure on a 737.

Still nothing.

A technician stood beside the power unit and pushed and pulled the lever harder and harder. He tried to get it to reverse, slamming it as hard as possible. "No piston reversal," the scribe wrote. The technician tried slamming it to the left. "No piston reversal." Then the group tested the yaw damper, an electrical device that made tiny adjustments to the rudder. "Passed. Passed. Passed. Passed. Passed. Passed. Passed. Passed. Passed."

The technicians removed the unique valve-within-a-valve and put it through a battery of tests. They probed it, flushed it, shined a light through it.

"Passed. Passed. Passed. No abnormalities found. No abnormalities. No anomalies. No anomalies. Normal wear pattern. No anomalies."

The power control unit was fine. The scribe ended the notes with a summary:

"The unit is capable of performing its intended functions as specified by Boeing. . . . Testing validated that the unit was incapable of uncommanded rudder reversal."

Phillips signed his name at the bottom of the page.

10

"WANNA PIECE OF THE PLANE?"

On a clear autumn day, Joan's friends and family came to the Saint Michael Catholic Church in Wheaton, Illinois. They lined up to sign the guest book beneath a photograph of Joan in her wedding dress. At the altar was another Joan photo, surrounded by candles and hundreds of flowers—carnations, dark red roses, and baby's breath. There was no casket. It was eight days after the crash and Joan's remains had not been identified.

Brett hugged people as they arrived. A few days earlier, he was worrying about mundane things like floor tile and getting to work on time. Now he had to deliver a eulogy for the woman he loved.

The priest who had counseled Brett and Joan before their marriage thanked everyone for coming and said that their presence in the church was a testament to the richness of Joan's life. "Death is not something to be explained," he said. "It is quite beyond our understanding."

Friends and relatives walked up for Holy Communion, knelt at the altar, and solemnly dipped pieces of bread into the wine. When Communion was over, Brett stood up and slowly walked to the podium, brushing his short auburn hair back from his forehead.

"When this occurred," he said tentatively, "it left me with a terrible, empty feeling inside. Going to Pittsburgh put this in perspective. There were one hundred thirty-two victims on that plane. Imagine all of us here tonight, multiplied by one hundred thirty-two, and you begin to get some scope of this plane crash. I would just like to offer my blessings for all of them as well, and all of the other people that are suffering through the same things."

He got a faint smile as he described Joan.

"Many times my friends and I would be sitting around the house watching a football game and Joan would be with us. She actually would watch the games. I'd say, 'Oh, no! Face mask!' and she'd say, 'Oh, no, that's encroachment.' She knew more about it than anybody I know. I always enjoyed going to work functions and parties with her. There would be an occasion or two where she would get up and speak. It was just remarkable to see the transformation from the woman I know that always played to my needs, to see her take on an executive's demeanor almost instantly, standing up and delivering a nice speech."

Brett said he had always been skeptical about religion and had never gone to church regularly, but he had been praying and watching for a sign from God that Joan had gone to heaven. He said he had received two since she died.

"I know I will get through this, as we all will, with difficulty and great pain," he said. "There's not much more I can say other than I loved her more than anything else in this world. . . . She's gone forever."

He paused. "I would just like to end with an Irish proverb that we both enjoyed:

May the road rise up to meet you
May the wind be always at your back
May the sun shine warm upon your face
May the rains fall softly on your fields
And until we meet again, may God hold you in the palm of His
    hand.

Three weeks after the crash, the Beaver County coroner still had not identified Joan's remains. Don Moore, a USAir pilot working in the morgue, had talked with Brett several times and sent him photographs of engagement rings found on several hands, but Joan's ring was not among them. Brett was in Melrose, Iowa, for another memorial service in honor of Joan when Moore called again. Brett was getting annoyed.

"Look," he told Moore, "it's got two triangular-shaped diamonds, two oval-shaped ones, and a marquise in the center. It's very unique." Moore promised to call back.

The phone rang again a few minutes later.

"Did she have a very thin, simple gold band for a wedding ring?" Moore asked.

"Yes."

"Mr. Van Bortel, we've identified your wife."

Joan's belongings arrived a few days later in a zippered white pouch that said, "Joan Van Bortel, aka Lahart, Gilbert Funeral Home." Inside were her checkbook, a phone bill, and some shattered credit cards. Her wedding and engagement rings were also included, although the engagement ring was missing a diamond. Brett hoped he could find the diamond at the crash site.

He had grown increasingly annoyed with USAir. The airline seemed to be backing away from its promises to pay some of his expenses. His family coordinator from the airline would not tell him who was sitting beside Joan. He knew she was in Seat 14E, but he wanted to know who was beside her when she died.

A week later, Brett drove to Hopewell Township to look for the missing diamond and to bury a gold brooch on the hill. The NTSB had finished its work there, and the Hopewell police were allowing family members to visit as long as they had an escort. Brett had arranged for Major Robert Pfeiffer of the Salvation Army to accompany him. He had met Pfeiffer when he visited Pittsburgh the weekend after Joan was killed. The night of the crash Pfeiffer had arrived on the hill very quickly, when the wreckage was still smoldering.

On this day they met in the parking lot at Green Garden Plaza and then proceeded up a driveway that led to the crash site. As they walked through the woods, Brett asked how it had looked immediately after the crash. Pfeiffer described the scene and the odd way some things were not damaged, such as an intact briefcase that was near others that had been shredded. As they walked through the trees toward the road, a boy came up to them.

"Hey, mister!" the boy said to Brett. "Wanna piece of the plane?" The boy was carrying slivers of aluminum.

"Goddamn NTSB," Brett muttered under his breath. Didn't they collect everything? How could they leave evidence behind? The wreckage was supposed to be safely inside a hangar somewhere. How could they solve the mystery if they didn't have all the puzzle pieces?

Brett stood there for a minute, trying to make sense of it all and then finally said sure, he'd take a piece. He looked around and found lots more. One was the size of his forearm. He walked down the road to the area where the nose had hit and knelt in the dirt. He was saying a quiet prayer when a woman tapped him on the shoulder.

"Your wife was on the flight?" the woman asked.

"Her name was Joan," said Brett.

The woman started crying. "My husband was . . ."

They hugged. The woman said she was Tina Connolly and that her husband was Robert Connolly, a financial consultant who was returning from a business meeting in Chicago. Brett explained that he had been trying to find out who sat beside Joan. Tina said she had been trying to find out the same thing—who had been sitting with her husband.

"Do you know where Joan sat?" she asked.

"Sure. Fourteen-E."

Her eyes widened and her chin quivered. She called to the man with her, who was her brother-in-law, and asked, "Do you remember where Bob sat?"

He shouted back, "Fourteen-F."

It took Brett a second to realize what that meant.

They stood there, staring at each other in amazement. Brett handed her a card about the scholarship fund he had set up in Joan's honor. The card had Joan's photograph on it. "Do you have a photo of your husband?" he asked.

"No," she said. She pointed to her brother-in-law. "But this is his identical twin."

They talked for forty-five minutes, discussing their experiences since the crash and plotting how they could compile a seating chart by talking with other families. Finally Tina left and Brett returned to bury the brooch.

Deciding to leave Brett alone, Pfeiffer climbed up an embankment to a wreath of silk roses. There were 132 roses on the wreath, each one tagged with a passenger's name.

Brett knelt in the dirt. He had not brought anything to dig with, and there were no sticks around, so he pulled out the piece of wreckage the kid had given him and used it to dig a little hole. He set the gold brooch inside and covered it up. He said a prayer and watched his tears fall to the dirt.

Just as Pfeiffer got to the wreath, a small gust of wind blew one of the roses to the ground.

He bent to pick it up and looked at the name on the tag: Joan Van Bortel. Pfeiffer came down to the road and handed the rose to Brett.

"Brett, it's a sign from God."

Brett filled his pockets with the wreckage of Flight 427, with squares of aluminum and a big bolt with a serial number on the side. There was so much debris that he could have filled a pickup truck. He took it back to his home in Illinois because he wanted proof that the NTSB had left evidence behind.

He had read in *Newsweek* that some crashes had been solved by evidence as tiny as a filament in a lightbulb. If that was the case, why didn't the NTSB take every piece? He called the *Newsweek* reporter who had written the story and described what he had found. The magazine ran a brief item about his findings in its Periscope section on November 7, 1994. The magazine quoted Brett as saying, "It's not like we had to excavate. The pieces were lying at our feet." An unnamed NTSB spokesperson was quoted as saying that investigators "took everything that was usable. We knew we were going to leave some pieces behind but we don't think they're of any consequence—what's left is for souvenir hunters." That infuriated Brett. Souvenir hunters?! Why would anyone want a piece of an airplane in which 132 people were killed?

A few days later, Mike Benson of the NTSB public affairs office called Brett and tried to explain further. Benson said the wreckage was left because investigators were getting a repetitive pattern. The smaller pieces were of no significance. Benson did ask him to send the bolt, however. But a few weeks later, Brett got a letter from the NTSB saying the bolt was of no consequence.

Brett had gone years without crying, but he was now breaking down every day. He went days without laughing. He thought it was strange, how he had always taken laughing for granted. But there was nothing in his life that was the least bit amusing. His first good laugh finally came when he noticed that Joan had left the price tag on the crystal frame he had taken to Pittsburgh. That was typical of her. She didn't worry about the small stuff.

It took several weeks to tell the bureaucracy that Joan had died. He found that the government, Joan's employer, Azko Nobel, and the credit card companies were sympathetic, but it was still painful to tell the story again and again. He increased his running. It was just about the only thing that provided any relief. He could put his mind on hold and focus on the pain in his legs instead of the one in his heart. He drove to the forest preserve where he and Joan had often hiked, and he ran sprints up and down Mount Trashmore. He put hundreds of miles on his mountain bike, riding a long trail that connected several nature preserves. He wore Joan's engagement ring on a chain around his neck now. It hurt sometimes when he rolled over in bed and the little prongs that had held the missing diamond dug into his chest, but Brett did not want to repair it. He wanted the ring left exactly as they had found it at the crash site.

Life felt meaningless to Brett, so he quit his job. He spent his days reading books about death and grief. He also visited the tiny Lisle library to read articles about the Flight 427 investigation, the FAA, and aviation safety. He learned that the FAA had a dual mandate—to regulate *and* promote aviation—simultaneously. What a conflict of interest! Critics said that the agency had a "tombstone mentality" and did not act until people had died.

His bitterness against USAir grew. He felt that the airline had been unresponsive to his requests about the Flight 427 seating chart. It wasn't until he and Tina Connolly had started compiling names that the airline decided to tell families who had been sitting beside their loved ones. He was also mad that the airline had reneged on some of its offers to pay for a church contribution and to fly friends and relatives to Joan's memorial service. The more he read about USAir's safety record, the more he disliked the company. The Pittsburgh crash was the airline's fifth fatal accident in five years. The FAA had beefed up its inspections of the company a couple of weeks before the Pittsburgh crash because of the other accidents. If Joan had known that, he doubted she would have flown USAir.

When Haueter's air traffic investigators checked radar tapes for other planes in the vicinity of Flight 427, they found that the closest was a Delta Air Lines 727 that was 4.2 miles ahead. That was more than a mile beyond the FAA's minimum spacing, so there was plenty of room between them. But when the air traffic group mapped the radar tracks of both planes, investigators discovered that the instant at which Flight 427 crossed the Delta jet's wake was the exact moment when things started to go wrong.

Big jets left wakes in the sky just as cruise ships left them in the water. Wakes were powerful spinning tubes of air the size of sewer pipes that kept spinning after a plane was miles away. They were invisible, but definitely noticeable. When a plane encountered a wake, it felt like a strong bump or it jostled the plane slightly to one side. Pilots could easily recover, but they might have to roll their wings level or adjust their heading.

Wake turbulence experts from NASA studied Flight 427's radar data and said there was plenty of evidence of a wake encounter. The calm weather on September 8 had been ideal for long-lasting wakes. The flight data recorder showed that the USAir plane's indicated airspeed jumped from 190 to 195 knots in one second. A big 737 cannot speed up that quickly, but that kind of change is often caused by an encounter with wake turbulence, which alters the flow into the airspeed sensors. But the NASA experts doubted that the wakes could have caused the accident. They might have rolled the plane slightly, but they weren't strong enough to flip the 737 upside down.

One of Haueter's bosses, Bud Laynor, was especially interested in the effects of a wake. As test after test cleared the plane's rudder valve, Laynor became more and more intrigued by the possibility that the wake had severely jostled Flight 427. But Haueter thought that was impossible.

"If that's so," he told Laynor, "why aren't we crashing them every day at National Airport?"

Still, Haueter acknowledged that it was possible that the wake had triggered some other kind of malfunction. The bumps may have activated the yaw damper, a device that made tiny adjustments to the rudder to make a flight smoother. If there was some kind of jam in the hydraulic valve, the rudder might have suddenly turned to one side—a problem known as a hardover.

Unfortunately, the government and the airline industry knew relatively little about wake turbulence. Engineers knew that wakes could be strong, but they didn't know how strong or how planes would behave when they crossed wakes at different angles. So Laynor began to push for a special flight test. He wanted to fly a 737 behind a 727 to re-create the conditions before the crash. Special equipment would measure the effects on the plane and show whether the wake could have flipped Flight 427 out of the sky.

The wreckage had been in the Pittsburgh hangar for seven weeks, the odor growing worse each day. It was a musty mixture of jet fuel, Clorox, and God-knew-what-else. The plane seemed to be decaying.

Dave Supplee, the USAir mechanic from Tampa, had been summoned back to the creaky hangar to rebuild the front sections of Ship 513. That wasn't unusual in a crash investigation, to put wreckage back together and look for patterns. But most accidents left large pieces that could easily be identified. That was not the case with Flight 427. There were thousands and thousands of pieces that were blackened, scarred, dented, flattened, and chipped. Putting them together was akin to working the world's most difficult jigsaw puzzle after the family dog had chewed up the pieces and they had been set on fire.

At Boeing's insistence, Haueter agreed that his structures team would reassemble the wheel wells, rudder cables, the leading edges of the wings, the floor beams, and the forward pressure bulkhead (the area just behind the nose of the plane). The group would examine the wreckage for evidence of birds, explosions, cracks, or fires. Haueter was still considering the possibility of a sudden failure of the rudder cables, a collapse of the floor beams or an explosion in the wheel well, a critical area where the landing gear came close to vital hydraulic lines, pumps, and cables.

Much of the USAir wreckage had been laid out on the hangar floor in the rough shape of the plane, but thousands of small, unidentified pieces had been tossed into three giant Dumpsters. The contents were poured onto the floor, and Supplee and other team members waded in.

"We're going Dumpster diving!" they exclaimed.

For two days they rummaged through the piles, looking for pieces they could recognize. Most were no bigger than a business card and were so badly charred that they looked like burned peanut brittle. The team started with the floor beams but found only 15 to 20 percent of them. So Supplee and his colleagues dove into the pile again.

Full-size blueprints had been spread out under Plexiglas so the team could locate each piece. Once Supplee found a piece that looked familiar, he searched for its location on the blueprint. He got a huge feeling of accomplishment when he was able to find the right spot. But the task was maddening. He kept asking, "Where's the rest of the plane?"

After several days, everyone realized the effort was impossible. There were huge gaps at the front of the plane. The wreckage had been so badly burned or so disfigured by the impact that there was no point in continuing. Supplee was discouraged. It seemed as if they had accomplished so little.

Still, the wreckage provided a few answers. Notches on the floor beams showed that the rudder cables had cut into them, which meant the cables were taut at impact and had not broken in midair. The burn pattern on the reassembled wreckage was random, which indicated that there was no fire or explosion until the plane hit the ground.

The partial reassembly of the front sections of the plane also allowed the structures team to look for evidence of birds. It was a long shot. Bird blood or feathers could have been washed away by the Clorox solution. But it was worth a check, especially around the slats, the movable panels on the leading edge of the wings. Boeing officials continued to suggest that a bird might have hit the slats right at the hinge. That could have caused the slat to pop up, making the plane roll to the left.

The bird scenario was one of a dwindling number of theories. Haueter was also interested in the gurgling sound that passenger Andrew McKenna had reported on the plane's previous flight from Charlotte to Chicago. The NTSB got an early explanation of the sound a few days after the crash, from a USAir pilot who sat in the cockpit on that flight and said his knee was leaning on a PA microphone button. But the issue had popped up again because of theories involving the electronics and equipment bay, where the computers and gyros were kept. There had been a few other incidents where water from the lavatory toilet had leaked into the bay and short-circuited the electronic hardware. The phenomenon was known as "blue water contamination," a nice way of referring to toilet water.

USAir had not been much help in pursuing the suspicions about the gurgling sound. The safety board had asked for a complete list of passenger names and phone numbers for the Charlotte-Chicago flight McKenna was on, but the airline had provided only thirty of them. To add to the mystery, Gerald Fox, USAir's O'Hare maintenance chief, did not tell investigators about the call he received about the strange noise until three months after the crash. The woman who reported the sound to Fox had never been found, nor had several other people who talked about the sound at Gate F6.

After dozens of phone calls, Haueter's investigators could track down only thirteen passengers, and only two of those had heard an unusual noise—McKenna and a woman who was sitting a few rows behind him. The woman did not think the sound was unusual, and she accepted the flight attendant's explanation that it came from the PA system. Haueter decided to put the gurgling inquiry to rest. The pilot's explanation about the noise was reasonable, and it seemed unlikely that the sound was important. If there had been leakage of blue water that caused a short circuit, it probably would not have made a sound and could not have moved the rudder as far as estimates said it had gone.

In the meantime, Flight 427's power control unit continued to pass every test. Samples of hydraulic fluid from the PCU, along with samples from other USAir, Southwest Airlines, and United Airlines planes, were sent out for testing. The safety board even spiked one sample with bleach to make sure technicians from Monsanto did not tamper with the results. The Flight 427 samples passed, although several of them were unusually dirty.

Technicians at Boeing tried a test using a typical rudder valve and hydraulic fluid so gummed up with particulates that it looked like Dijon mustard. The contamination was much worse than anything the investigators had found on the plane, but they wanted to see how much the valve could endure before it malfunctioned. The fluid was so thick that it kept destroying Boeing's hydraulic pumps. But in test after test, for thirty hours, the equivalent of stomping on the rudder pedal five thousand times, the valve worked fine.

The bird theory was still alive, mostly because Boeing was so insistent about it. Members of the structures group were back at the Pittsburgh hangar, where the windows had been covered with black paper. They shut off the lights, donned strange-looking orange sunglasses, and shone a black light along the wreckage. The sunglasses were supposed to make it easier to see the faint glow from bird remains. It was a bizarre scene: people wearing weird sunglasses walking slowly around the wreckage, waving a black light the size of a baseball bat.

On the number three leading edge slat, they saw a faint glow. They brought the light closer. It wasn't bright, but it was definitely stronger than the surrounding area. They switched on the hangar lights and examined the spot carefully. It was a small clump of material that looked like dirt or mud. It wasn't bird proof by any means, but it was worth testing.

It was time to call Roxie Laybourne, the world's premier expert on feathers.

Laybourne, who was eighty-four years old, was a legend among crash investigators. She was a tiny woman with short gray hair and thick eyeglasses who worked at the Smithsonian Institution's National Museum of Natural History. She wore a white lab coat and Reeboks, and she walked hunched over because of back and neck problems. She had begun studying feathers in 1960 after a Lockheed Electra collided with a flock of starlings in Boston, killing sixty people. Suddenly the government cared a lot about birds.

Identifying them was an odd but necessary task in aviation. Engine and airplane manufacturers needed to know how well their products could take a direct hit from a laughing gull, a Cuban whistling duck, or—their biggest fear—a Canada goose. Once they got Laybourne's reports, they knew how much damage different birds could cause, which enabled them to improve their products to better withstand a bird strike.

Examining feathers under her microscope, Laybourne had discovered unique patterns in the stringy fragments of feathers called downy barbules. Some barbules had little triangles, others had rings. She found similarities between birds from the same family, even those that lived thousands of miles apart. Pigeons and doves were easiest to identify because the nodes on their barbules looked like crocuses. The hardest were songbirds because their nodes all looked alike.

With tiny feather fragments—some no bigger than bread crumbs—Laybourne solved thousands of bird strike cases. In one, she used a fragment of down from a pilot's shoulder patch to identify the herring gull that broke through the canopy of a Harrier military jet. She also used her skills to help Customs identify birds smuggled into the United States, and she once helped the FBI catch a murder suspect.

Laybourne's office at the museum was a reflection of her unusual occupation. Her bookshelf included *Birds of Nepal* and *Raptors*. On her door were a poster from the Israeli Air Force warning pilots to watch out for birds and a *Far Side* cartoon that showed Santa Claus and his reindeer splattered on the nose of a jumbo jet. She was philosophical about the conflict in the skies. "As long as you have man and birds flying," she said, "you have the potential for problems."

Cindy Keegan, the head of the airplane structures group for the Flight 427 investigation, brought the suspicious clump to the museum. But when Laybourne saw the tiny brown sample, she doubted she could be much help. It was smaller than a dime, a mixture of sand and dirt but nothing that even remotely looked like a feather. She eyeballed it for a moment and then examined it under the magnifying glass she wore around her neck. Nothing. She pulled off the cover on the microscope, put a sample on a slide, and took a closer look. Still nothing.

The next day, two other bird strike experts happened to stop by her office. One, Major Ron Merritt, headed the Air Force's famous BASH Team. The other, Eugene LeBoeuf, was the FAA's chief bird scientist. After examining the sample, they agreed that it was only dirt and vegetation.

Laybourne phoned the NTSB. "There's no bird material here," she said.

11

BACKDRIVE

The data recorder on Flight 427 was like an eyewitness to the crash with one eye closed. It took only thirteen measurements of the flight, which provided an incomplete picture of what happened. It saw the plane yaw to the left and then roll out of the sky, but it didn't see why. That meant the NTSB and Boeing had to make educated guesses about what had happened. Investigators were pretty sure the sudden yaw was caused by the rudder. No other flight control on a 737 could cause that kind of movement. But they did not know whether the rudder moved because the pilots stomped on the pedals or because something went wrong in the hydraulic system.

That had become the central question in the investigation: Was the crash caused by man or machine? As other theories were ruled out, Boeing and ALPA began to differ about why the rudder had moved. Boeing kept raising the possibility that the pilots had mistakenly stomped on the pedal, whereas ALPA kept suggesting the rudder system had malfunctioned.

The small number of measurements on flight data recorders was a function of money. If it were up to the NTSB, every airplane in the United States would be required to have 300-parameter boxes immediately. But the airlines balked at that request because it cost tens of thousands of dollars to equip each plane with the more advanced recorders. It wasn't that the boxes were any more expensive. The recorder on Flight 427 could easily have handled dozens of additional parameters. It was the multitude of new sensors and wiring that ran up the cost. Because the 737 was a relatively basic airplane with steel cables that moved the flight controls, it was costly to install sensors.

A newer, fly-by-wire plane could be equipped with a sophisticated flight recorder more cheaply because the sensors and computers were already there.

Boeing assigned some of its smartest aerodynamic engineers to perform a kinematic analysis of the crash. Their goal was to take the undisputed facts from the thirteen measurements in the flight recorder and come up with estimates for things the box did not record, such as movement of the wheel and the rudder. The job was even tougher because the plane had been bounced around by the wake from the Delta 727. There was no doubt that the wake had jostled the USAir plane, but it was up to the engineers to figure out which movements were caused by the wake and which were caused by the ailerons and rudder.

The debate about why the rudder moved was fueled by the fact that 737 pilots rarely used it in flight. Pilots of smaller planes had to use the rudder during turns, but the aerodynamics of the 737 were different, and there was no need to use the rudder during a turn. Pilots used it primarily for landings in strong crosswinds and in the rare case of engine failure.

To help sort out what had happened to Flight 427's rudder, the NTSB and Boeing decided to try a technique known as a backdrive. Engineers took the numbers from the flight data recorder and fed them into the computer that ran Boeing's M-Cab flight simulator. The *M* in "M-Cab" stood for "multi-purpose," which meant the simulator was a chameleon. It could be programmed to perform like any Boeing jet. M-Cab was usually used for dull engineering tests, but this time it would be used to re-create the crash and see how 132 people had died.

It was a ghost flight. The plane would roll and twist the same way it had on September 8, as if the pilots were still at the controls. The instruments would spin the same way they had in Ship 513, and the control column would move back and forth as if Emmett were still pulling on it, trying to prevent the crash.

In a cavernous room three stories high, M-Cab stood like a long-legged elephant with a Boeing logo on the side. Its legs resembled huge shock absorbers, allowing the cab to rock back and forth so pilots believed they were flying. It was built to fool the pilots, using gravity, motion, sound, and computer displays to give a realistic sensation of flight. There was a sense of danger when M-Cab was about to start, as if the big beast might go crazy, fall off its legs, and crash to the ground. That was especially true with the Flight 427 work because in order to simulate the deadly plunge of the plane, Boeing had to push back the mechanical safety stops that limited how far M-Cab could go.

The brains of the beast were in a computer room next door, where technicians controlled the cab. Computers the size of refrigerators were lined up at one end of the room. At the other end, technicians and engineers could watch the white cab and talk to the pilots through an intercom. Pilots referred to the Flight 427 simulation as "rock and roll." They often began the rides by saying, "Let 'er rip."

Haueter walked across the ramp to the white cab and climbed into the mock cockpit. He was curious to find out what the ride would show. He had studied charts about what had happened to Flight 427 dozens of times, but it would be much better to feel what the pilots had encountered instead of looking at lines on a piece of paper.

He knew the pilots had made a crucial mistake. They pulled back on the control column at a critical time, and that made the plane stall. Had they just pushed the stick forward and let the plane lose some altitude, they probably could have prevented the crash. Also, the pilots had not acted quickly enough to turn the wheel completely to the right, which might have countered the effects of a rudder hardover, regardless of whether it was caused by man or machine.

Haueter was not shy about blaming pilots. He had gone toe to toe with ALPA on the 1988 crash of Delta Flight 1141 in Dallas–Fort Worth, and he'd been amazed at how ridiculous the union had been, claiming there was some kind of malfunction of the plane that caused it to crash a few seconds after takeoff. But Haueter and the other investigators were convinced that the pilots had simply forgotten to set the flaps, the panels on the wings that gave the plane extra lift so it could get off the runway. Haueter had also blamed pilots for a crash in Hawaii—the guy flew right into a cliff—and one in Alabama in which the pilots got hopelessly lost. Yet he was still skeptical about whether Emmett or Germano would have stomped on the pedal and held it there. Good pilots—especially ones with thousands of hours like Emmett and Germano—just didn't do that.

Haueter took the right seat, where Emmett had sat. He reached below to a lever and adjusted the seat until his feet could reach the rudder pedals. He was about to see how 132 people had died, but he would not get choked up about it. He coped with the gruesome side of his job by building a big emotional wall. The facts got in, but the horror stayed out. Still, he expected it would be a bumpy ride and figured he would get a good jolt from the G-forces.

The cockpit looked the same way it had to Emmett and Germano. The altimeter read 6,360 feet. The airspeed indicator read 190 knots. Looking through the cockpit window, Haueter could see blue and green squares that were supposed to represent Hopewell Township and the surrounding area. Jim Kerrigan, a Boeing engineer riding with Haueter, told the M-Cab technicians that they were ready. Flight 427 began again.

M-Cab descended from 6,300 feet and banked slightly to the left. Haueter heard the steady *click-click-click* of the rudder trim. It felt like a routine turn. Boeing did not have a copy of the cockpit tape, but Haueter had heard it so many times he knew what the pilots would say at each moment. *Oh, yeah, I see zuh Jetstream. Sheeez. Thump. Whoa.* The cab jerked to the left and then back again, rattling the metal seats and panels and tossing Haueter left and right. *Hang on, hang on.* Another jerk to the left and then the *whoop-whoop-whoop* of the autopilot warning horn. Haueter had been watching the computer image of a blue sky and hazy horizon out the cockpit window, but suddenly all he could see was the green and blue squares of the ground, looming closer. *Hang on. Ohhh, shiiiiit. What the hell is this!!?* He had his hands lightly on the control column and felt it come back, as if the phantom pilots were trying to pull the nose up to avoid certain death. He felt the stickshaker go off like a jackhammer, warning that the plane was stalling. He was almost face down now, watching the ground spin closer and closer. *Oh shit! Pull!* The jackhammer continued *rat-a-tat-tat-rat-a-tat* all the way down, until the ground filled the windscreen *God! Pulllllllll!* and the cab jerked to a stop. *Noooo.*

As violent as the end was, Haueter was surprised that the initial yaw and roll were so smooth. He had expected to get bounced around more severely before he was rolled to the left. But the whole thing happened seamlessly and fast. They were making a routine turn like they had thousands of times, and twenty-eight seconds later they were dead.

Haueter flew it again. He let the machine jerk him back and forth as he looked for some sort of cue that would have triggered Emmett or Germano to stomp on the rudder pedal. A 737 pilot might use the rudder in a crisis, to quickly counteract a roll. But in that case a pilot most likely would step quickly on the pedal and let it out—not hold it down all the way to his death. As the phantom pilot again pulled the stick back, Haueter searched for a reason they might have held the rudder.

He couldn't think of one.

He flew it ten more times. The more he flew, the more unlikely it seemed to him that Emmett or Germano had stomped on the rudder pedal. There were a few bumps from the wake, but nothing that would make them press the rudder all the way. Instead, it felt like an aerobatic move called a spin entry. A stunt pilot deliberately held the rudder in and pulled back on the stick so the plane would flip out of the sky and corkscrew toward the ground.

Haueter unsnapped his seat belt and climbed out of the chair. As he walked out of the simulator, he was even more convinced. Nothing had happened to Flight 427 that would have made a seasoned pilot stomp on the pedal.

Within days of the crash, families of the victims of Flight 427 were inundated with mailings from lawyers.

"It is most difficult to intrude into your life at this time, but you do need help, not only spiritual, but legal as well," wrote the Cleveland firm of Miller, Stillman, and Bartel. The letter promised the firm would "only accept five families from any one disaster, as each family is entitled to our utmost time and individual attention." Included was an ad that listed the death and injury cases the firm had handled—planes, trains, automobiles, even "slips and falls."

Brett received mailings from about thirty lawyers. He piled them on his office floor in a two-foot stack. A few were downright tacky. One lawyer sent a refrigerator magnet. Another sent a videotape.

"First I want to offer my sincere condolences," Houston lawyer John O'Quinn said on the five-minute tape. "I realize that sudden losses such as this are very difficult to deal with, and unfortunately at this time your mind is probably not prepared to make certain important decisions which unfortunately must be made." The tape apparently had been recorded specifically for families of the Flight 427 victims. "I assure you that USAir already has its attorneys working on the case . . . in the best interests of USAir and not necessarily in your best interests." Newspaper headlines flashed on the screen showing examples of his firm's victories, followed by a toll-free phone number that families could call.

For a trial lawyer, a plane crash case is a sure thing. It's not a question of *whether* a plaintiff such as Brett will win, but *how much* he'll win. As in other types of personal injury cases, the lawyers usually work on contingency, getting paid only when their clients win. (Which means that, in airline cases, the lawyers always get paid.) The contingency fees are typically 15 to 25 percent, although some lawyers get as much as 35 percent.

The settlements are based on the odd practice of determining the price of a human life. To people outside the legal system, it seems cold and insensitive to try to translate a human life into dollars and cents. How could USAir and Boeing possibly repay Brett for the love he shared with Joan? How could they put a dollar value on her life? But by filing a lawsuit against them, Brett was seeking money as compensation for her death, which meant the court would have to determine what Joan was worth.

Much of the money came not from the airlines and aircraft manufacturers but from their insurance companies, which meant that lawsuits from a crash had little direct effect on an airline's bottom line. USAir had $850 million in coverage spread among seven companies in the United States, England, and France. The largest, Associated Aviation Underwriters of Short Hills, New Jersey, had 30 percent of the coverage and would be the lead insurer to handle claims on the policy and negotiations with the victims' lawyers. It's likely that those seven companies had gone to the insurance markets in London and gotten "reinsurance" on the policy, which spread the risk among more investors. USAir has never disclosed how much it paid for this coverage, but the trade journal *National Underwriter* reported that insurers canceled and renegotiated USAir's coverage after the Charlotte crash—one month before the Hopewell accident—to raise rates or extend the payment schedule. The journal referred to USAir as the "hapless carrier."

Boeing had its own coverage, probably with some of the same companies that USAir did, but Boeing did not publicly discuss details of its insurance and, unlike USAir, it was not required to file a statement of coverage with the government. USAir and Boeing also would not say how they would divide their liability for the crash. In previous cases, airlines and aircraft manufacturers have often started with a certain split, say fifty-fifty, and then revised the proportion as it became clear which company was more at fault.

Although many people criticized lawyers for soliciting clients after a disaster, Brett found that the mailings were helpful in that they explained the process of a wrongful death case. He didn't know much about the legal system, and the information that he received answered many questions. But ultimately he decided to go with Corboy Demetrio Clifford, a Chicago firm that had not sent him a single piece of mail. The firm, which had been recommended by his father's lawyer, was so well known for aviation cases that it did not need to solicit clients. Its offices took up two floors in a tower at 33 North Dearborn Street, across from the Cook County court building. The walls and cabinets in the law firm were polished mahogany, the desks were marble. The firm won so much for its aviation cases that it could afford a lot of mahogany.

The firm was headed by one of the city's most famous trial lawyers, a white-haired legend named Philip Corboy. He was famous for winning millions for clients—$25 million for the family of a passenger killed in the 1989 United Airlines crash in Sioux City, Iowa, and nearly three hundred other verdicts and settlements for $1 million or more. A competitor called him "the Hoover vacuum of the personal injury business." Brett met Corboy briefly, but ended up dealing primarily with partners Robert Clifford and Michael Demetrio.

They didn't waste a minute filing Brett's suit. Within hours after he signed the contract to hire them on October 12, his suit was hand-carried across Dearborn Street to the clerk's office in Cook County Circuit Court. The decision to file there was a strategic one. Cook County juries were famous for giving big awards, and Corboy Demetrio Clifford had considerable clout there. The Chicago court was sure to be a better venue than Pittsburgh, where USAir was a major employer and was regarded as the hometown airline.

Brett said his decision to sue USAir and Boeing was more about justice than about money. He wasn't looking to get rich. He would get plenty of money from life insurance policies and workmen's compensation for Joan. What he wanted was to get revenge against the companies he believed had killed her.

And so Brett's revenge became Case No. 94 L 12916, *Van Bortel v. USAir Inc., the Boeing Company and Gerald E. Fox.* By naming Fox, the airline's Chicago maintenance foreman, Brett stood a better chance of keeping the case in Cook County rather than having it transferred to Pittsburgh, where many other cases were being filed. The lawsuit mentioned a host of possibilities that could have caused the crash: inadequate maintenance, the thrust reversers, the rudder system, the electrical control system. It said the airline "allowed USAir Flight 427 to crash into the ground." In the complaints against Boeing, the suit said the 737 was "unreasonably dangerous" because its rudder system or thrust reverser did not respond properly or because it had a faulty anti-collision warning device or had an engine that fell off. The suit said Fox "carelessly and negligently failed to properly conduct the preflight maintenance and/or inspection of Flight 427." The scattershot nature of the complaint was standard practice for trial lawyers. It was too early to know the cause of the crash, so they made a broad range of accusations. The suit said Brett "has sustained pecuniary loss and damage, including loss of society and companionship, love and affection, as a result of her death."

The accident record for 737s was better than those of many other jets. But the record carried a disturbing asterisk that haunted the NTSB investigators: the Colorado Springs crash. It was one of only four cases that the NTSB had never solved.

United Airlines Flight 585 was supposed to be a twenty-minute hop from Denver to Colorado Springs. Skies were clear, but the Colorado Springs airport was having gusty winds.

"Nice lookin' day," Captain Hal Green said to First Officer Trish Eidson as the plane taxied away from the gate. "Hard to believe the skies are unfriendly." As they headed toward the runway, one of the pilots whistled.

The two pilots talked about the danger of rotors, the spiraling winds that came off the mountains. Green said he knew a pilot who had flown through one. Eidson said she knew they were dangerous and that they "could tear a wing off."

The flight was so short that as soon as they climbed to 11,000 feet they had to begin preparations for landing. Eidson tuned to the latest weather report for Colorado Springs, which said the airport had a low-level wind shear advisory with winds gusting to 40 knots. An air traffic controller gave her directions to descend and turned them toward the airport.

"Cleared to land," the controller said. "Wind three-two-zero at one-six, gust two-niner."

"Okay, we're cleared to land three-five, United 585," Eidson replied. "Getting any reports lately of loss or gain of airspeed?" She wanted to know if planes were encountering trouble because of the turbulence.

"The last air carrier was the one that reported that, a 737," the tower said.

"Could you repeat it please?"

"Yes, ma'am, at five hundred feet, a 737-300 series reported a five . . . correction, a fifteen-knot loss. At four hundred feet, plus fifteen knots and at one hundred fifty feet, plus twenty knots."

"Sounds adventurous," Eidson replied. "United 585, thank you."

She then said to Green, "I'll watch that airspeed gauge like it's my mom's last minute."

They were parallel to the runway, preparing to turn to their base leg and then make their final descent. "Okay, start around there now and wheels down final," Green said.

"Gear is down, three green," Eidson said. "Speed brakes armed, green light, flaps are five, green light, hydraulic brake pressures are normal, final descent checklist complete."

As they descended toward the runway, their airspeed changed suddenly. "We had a ten-knot change there," Eidson said.

"Yeah, I know. Awful lot of power to hold that airspeed," Green said.

It went up ten more. "Wow," said Eidson.

Twenty seconds later, she said, "Oh God."

The plane was rolling sharply to the right.

"Oh!"

"Oh!!!!"

The pilots tried to adjust the flaps, apparently trying to abort the landing and go around.

"No!"

"Oh my God!" Eidson screamed. "Oh my God!"

"Oh no!"

The plane crashed nose first into a park three miles from the airport and exploded in a fireball. All twenty-five people on board were killed. A witness said the aircraft had looked like it was on a dive-bombing mission.

The NTSB explored dozens of theories about the engines and the plane's rudder. Many leads went nowhere. Investigators studied sound frequencies on the cockpit tape to explore whether the engines were operating normally. But gaps in the sounds prevented them from drawing conclusions. They searched for an elderly couple who reportedly had been splattered with a bad-smelling liquid as the plane flew overhead, but the couple could not be found. As in the Hopewell crash, the flight recorder was of limited help. It had only five parameters and because the plane was in turbulence, it was difficult for NTSB engineers to determine whether the movements of the plane were caused by the weather or by its flight controls.

Much of the evidence pointed to the rudder and the likelihood of a slow-moving swing to the right. Tests in flight simulators showed that could have caused the accident. Investigators discovered that the plane's standby rudder actuator—a backup device that would be used only if other hydraulic systems failed—had spots where metal had scraped against metal. And in the week before the crash, pilots of the same airplane had twice experienced sudden yaws to the right, problems that were attributed to minor malfunctions in other rudder components. But tests on the power control unit were inconclusive. Many PCU parts were badly burned in the crash, which raised doubts about the validity of the tests.

Engineers found that a jam in the soda can valve could not have created the slow turn by the rudder that would have been needed to match the flight recorder. Also, internal stops in the valve would have prevented other malfunctions from moving the rudder very far. And even if there had been a malfunction, Green and Eidson should have been able to counter the rudder easily with other controls.

Weather experts studied satellite pictures of Colorado on the day of the crash, looking for evidence of a powerful rotor, the horizontal tornadoes that Green had mentioned before takeoff. A glider instructor said he saw a rotor hit the ground the day of the accident, blowing off tree branches and damaging car hoods. Pilots in the area had reported turbulence and downdrafts. A 737 pilot reported a "good sinker." But another 737 pilot described it as a normal windy day in Colorado. Meteorologists said rotors usually made a loud roaring sound, but witnesses to the crash said they didn't hear anything out of the ordinary.

When the NTSB asked the parties to submit their ideas, Boeing and ALPA offered starkly different explanations for the crash. Boeing blamed the wind. The company said a powerful rotor struck the plane at low altitude and "did not allow sufficient height for recovery." But the pilots union said the evidence was inconclusive, that there were strong indications of heavy winds, but the possibility of mechanical problems could not be ruled out.

Ultimately, the safety board could not reach a conclusion. After long discussions, the board decided it did not have enough for either theory. There was strong evidence implicating the rotor, but not enough to prove that it caused the crash. There was evidence that the rudder may have malfunctioned, but that should not have been enough to flip the plane. When the NTSB published its final report on the crash, the cover carried big blue embarrassing words: UNITED AIRLINES 585 — UNCONTROLLED COLLISION WITH TERRAIN FOR UNDETERMINED REASONS.

Safety board investigators pledged to keep looking for a cause. "We're worried," NTSB official Ron Schleede told the *Washington Post,* "that we may have overlooked something."

The Colorado Springs crash had haunted the safety board investigators ever since. To some, it was an embarrassing admission of defeat. UNDETER-MINED REASONS. Those were painful words, an acknowledgment that the NTSB wasn't up to the job. But to others, that was the responsible approach: to lay out the evidence and admit that it wasn't conclusive.

Flight controls specialist Greg Phillips worked on the Colorado Springs investigation and was involved in many of the tests on the valve, but he believed there wasn't enough evidence to blame the airplane. Still, he kept a running list of 737 rudder incidents in his file drawer. Three months before the USAir crash, he met with Boeing officials about several rudder-related incidents and questioned whether Boeing had adequately informed airlines about fixing a rudder system component. When he flew to Pittsburgh, he didn't need to bring the charts from the Colorado Springs flight data recorder. He had them memorized.

When the Colorado Springs and Hopewell crashes were reduced to lines on a graph, there were a few similarities. The charts for airspeed and altitude showed the same basic arches. At first glance, the flights looked like mirror images of each other, the United plane rolling right, the USAir plane rolling left. The cockpit tapes showed that both crews were caught by surprise.

But there were major differences. The United plane was at 1,000 feet, the USAir plane at 6,000. The United plane was going much slower, using more flaps on its wings. The United plane was in windy weather, the USAir plane was in calm. Boeing discounted the similarities, saying that the company still believed the Colorado Springs crash was caused by the rotor. Jean McGrew, Boeing's chief 737 engineer, said the only similarity between the two accidents "was that they are both 737s."

Al Dickinson, the lead investigator in Colorado Springs, was a friend of Haueter's. Dickinson was smart, a good investigator. But in the years since the crash, the guys in the office had come to look at Dickinson differently. He would always have an albatross around his neck because he had investigated the big one and come up short. Haueter reassured Dickinson that his work on United 585 had formed the foundation for everything Haueter did on the USAir crash. The NTSB was much farther along because of Dickinson's persistence, Haueter told him. But Haueter was determined that his crash would not be unsolved. He vowed that the words *UNDETERMINED REASONS* would not appear in his report.

On a Saturday in January 1995, Haueter drove to his office in L'Enfant Plaza to finish some paperwork in preparation for the public hearing on the crash. The office was like a ghost town, dark and lonely on weekends. Haueter didn't like the silence, but he could get a lot of work done with no interruptions. He worked for several hours and then put on his jacket to leave. As he walked out, he went by a conference room and noticed some posters on the walls. He stepped inside.

The posters were tacked up on three walls of the cramped room. Each of them represented a second or half-second increment of Flight 427. The thirty-two posters showed the view that Germano would have had from the captain's seat, coupled with Boeing's estimates of how the wheel and rudder pedals had moved. Boeing had used footprints to show how the rudder moved, a subtle dig at ALPA that suggested the pilots moved the rudder with their feet.

Haueter had seen the same data hundreds of times—in charts, video animations, computer spreadsheets, and M-Cab—but this suddenly brought everything together. He walked along the wall, studying where the rudder pedals had supposedly moved. The USAir 737 had been jostled by the wake of the Delta plane and then rolled to the left. That made sense. Then one of the pilots tried to stop the roll by turning the wheel to the right. That made sense. Then the left rudder pedal went in briefly and came out. That made sense. One of the pilots probably tried to slow the plane as it was rolling back to the right. But then the left pedal went in again, almost all the way.

That made NO sense.

Why would a pilot keep applying left rudder when the plane already was rolling left? That would be like a driver realizing his car was veering toward a cliff and then turning the wheel to go over the cliff sooner.

Haueter kept studying the drawings and the rudder estimates, checking when the rudder came in and when the wheel moved. It all seemed out of sync. Then a light went on in his head. What if the rudder drawings were reversed? Instead of pushing on the left pedal, maybe the pilot was pushing on the right pedal, but there was some malfunction in the power control unit that caused a reversal and made the rudder go the wrong way.

Suddenly it all made sense. The pilot had been stomping on the right pedal, trying to stop the big plane from rolling left, but the rudder wasn't responding. In fact, he was actually pushing the rudder the opposite way, causing the plane to roll out of control. That would match the grunts and confusion on the cockpit tape. When Germano said, "What the hell is this?" maybe he was referring to rudder pedals that were not responding the way they should.

Haueter left the room, took the elevator downstairs, and went outside. He walked down the stairs beside the subway station and took a shortcut beneath the Department of Housing and Urban Development, a beehive-shaped building beside L'Enfant Plaza. As he crossed D Street in the brisk winter weather, he thought about what he had seen.

He didn't know what could have made the rudder reverse. Phillips had subjected the power control unit to every malfunction they could dream up and it had withstood every one. And there were no marks showing that the valve had jammed or reversed.

He jaywalked across Seventh Street and walked beneath the railroad tracks, where a platoon of pigeons had found shelter from the cold January weather. He turned right on Maryland Avenue, saw the dome of the U.S. Capitol looming in front of him, and then crossed Independence Avenue to the Smithsonian Air and Space Museum. This was Haueter's secret retreat, the place where he went to escape his worries. He occasionally sneaked over on workdays when he needed to clear his head and get inspiration from the world's greatest collection of historical planes. He was always impressed at how much progress had occurred in such a short time, from the Wright brothers to the moon landing in one lifetime. There was still a little boy inside Haueter who marveled at the beautiful planes. Even if he couldn't fly them, he would love to just sit in them.

He walked in the main doors, beneath the *Voyager* plane that had circled the globe, and headed for the west end of the building. He walked up the stairs to the second floor beneath a silver DC-3. FLY EASTERN AIR LINES, THE GREAT SILVER FLEET, it said on the side. He browsed in the World War I gallery and then went to the Pioneers of Flight room. He stopped in front of the Lockheed Sirius, a red two-seater with big silver pontoons that Charles and Anne Morrow Lindbergh had flown to the Far East. Charles Lindbergh had always been one of Haueter's heroes. A picture of the *Spirit of St. Louis* hung on his office wall.

As Haueter admired the plane, his thoughts circled back to the investigation. He felt more confident than ever that the rudder had reversed. It made perfect sense. Now he just had to figure out why.

GREMLIN

It didn't take long for nasty jokes about USAir to show up on the Internet. One list suggested new advertising slogans for the airline: "When you just can't wait for the world to come to you"; "Complimentary champagne during free-fall"; and "The kids will love our inflatable slides." Another list claimed to have the cockpit tape with the final words of Flight 427's pilots:

"Pittsburgh will be our *final* destination."

"Let's see if this baby can do 300!"

"Oh stewardess! Oh yes! Oh yes! Oh no . . . "

"Assume the position."

USAir was an easy target for wisecracks not just because of the five crashes but also because it had never been considered a big-time airline. American and United had been major carriers since the early days of commercial aviation, serving big cities across the country. But USAir's roots were a hodgepodge of regional airlines that served the blue-collar towns the big airlines didn't care about.

USAir began in 1939 as a quirky little company called All American Aviation that went dive-bombing for airmail packages in the Allegheny Mountains. Most of the towns were too small or remote to justify full air mail service, so All American's planes dove toward a contraption that looked like a football goalpost, dropped a container of mail, and then snagged the outgoing package with a hook. A clerk inside the plane reeled in the package as they flew to the next town.

The years when All American was grabbing packages in Punxsutawney and Oil City and Slippery Rock helped it develop an image of a gritty regional airline serving the Rust Belt. And while All American was becoming famous for the drudgery of picking up envelopes, carriers such as Pan Am and United were flying wealthy business travelers to the nation's biggest cities, earning reputations for pampering them with luxurious service.

By 1953, All American had changed its name to Allegheny Airlines and had begun carrying passengers. It had a fleet of thirteen DC-3s serving such cities as Washington, Baltimore, Philadelphia, New York, and Newark. Over the next twenty-five years, Allegheny grew steadily by merging with Lake Central and Mohawk to become the main regional carrier in the Northeast. But it still was not in the same league as the big guys. Passengers complained about mediocre service and dubbed it "Agony Airlines."

With the government tightly regulating the airline industry, Allegheny was limited to routes in the Northeast. But the company had a solid market. The government was generous in setting fares, and most airlines could make a profit with planes that were half empty. Then came the revolution, the deregulation of the airline industry in 1978. Allegheny was freed to compete for national routes. It began flying to Phoenix, Tucson, New Orleans, and Raleigh and changed its name to USAir to get rid of the small-time image. Because it did not have the critical mass that the big carriers had, USAir had to buy other airlines to grow. The company's chairman, Edwin I. Colodny, was a visionary who understood that under deregulation an airline had to be big to survive. He bought Pacific Southwest Airlines in 1987 to get a toehold in California and then merged with North Carolina–based Piedmont Airlines in 1989.

But those mergers gave USAir indigestion that lasted for years. Colodny misjudged the difficulty of combining three very different companies, which created a mishmash of cultures, airplane types, and offices that took years to sort out. The Piedmont employees resented the swiftness of the merger and the way their friendly Southern company had been swallowed overnight by a bunch of Yankees. In defiance, many pilots continued to wear their Piedmont wings and drew graffiti of the Piedmont logo in USAir cockpits. Likewise, USAir was unprepared for the brutal competition on the California routes, where Southwest Airlines and America West were slashing fares. USAir soon made an embarrassing retreat from California, conceding most intrastate routes to its competitors.

Airlines have always been especially vulnerable to changes in the economy. When the economy is bad, people don't fly. And once it improves, people are slow to resume flying. The airlines are both a leading economic indicator and a lagging follower. They feel a recession sooner, and it takes them longer to recover. But USAir felt the dips even more than its competitors did because it had such high costs and the mergers made the company so inefficient. It had an expensive mix of nine different airplane types, which meant more spare parts, more training, and more maintenance hangars. Even before the mergers, USAir had higher costs because it was heavily unionized and it operated in the most expensive region in the country. The mergers added an extra layer of complications. Other airlines had two or three hubs spaced across the United States, but suddenly USAir had five clustered in the East. The company had too many reservation centers, too many crew bases, too many hangars. It was a recipe for red ink.

USAir had lost money every year since the 1989 Piedmont merger, for a total loss of $2.5 billion. The company chairman, Seth Schofield, had taken steps to shrink the fleet and close hubs, but the airline was still so inefficient that its operating costs were far and away the highest in the industry. The back-to-back crashes in Charlotte and Hopewell dealt a painful double whammy, scaring away thousands of customers. The airline lost $10 million in bookings because of the Charlotte crash and another $30 million in just three weeks after the Hopewell crash. Its stock price had fallen by 35 percent two months after the crash, down to just $4.25 a share. Some airline analysts were predicting that the company was headed for bankruptcy court.

USAir had tried to rebuild its image. Schofield and senior vice presidents held meetings with big corporate customers to reassure them about the airline's financial health and commitment to safety. In a letter to more than a million of the airline's frequent fliers, Schofield wrote: "I want to reassure you that your confidence in the operational integrity of USAir is of paramount importance to us."

By mid-November, it seemed the campaign was paying off. Bookings had rebounded almost to normal levels. It appeared the airline had turned the corner.

Then came the story in the *New York Times*.

TROUBLES AT USAIR: COINCIDENCE OR MORE? asked the front-page headline on Sunday, November 13, 1994. The story painted a picture of sloppiness and cost cutting at the beleaguered airline, where pilots departed without enough fuel and where a maintenance supervisor tried to save money by letting a plane fly with an inoperative stall-warning system. The *Times* followed the classic formula for a blockbuster investigative story—scary anecdotes, lots of statistics, quotes from worried experts, and weak-sounding denials from the airline. The newspaper said it had obtained "a previously undisclosed report" from the FAA and quoted federal officials who "spoke on the condition that their names be withheld." A front-page chart showed that the risk of dying on USAir was 5 deaths per 10 million passengers, compared with 1.5 deaths for other major airlines. Once again the USAir name was being linked with death.

The story ricocheted around the country. Newspapers that subscribed to the New York Times News Service published shorter versions of it the same day. It was picked up by the Associated Press and became the lead item on TV newscasts in many cities. Sunday is the slowest news day of the week, so producers were happy to get a story with some pizzazz. "Shocking discoveries raise questions about safety procedures at USAir," said a Detroit news anchor. "USAir is under attack for its safety record," said one in Philadelphia. The *New York Post* said, "USAir's future is in doubt after revelations that it skimped on safety precautions to cut costs, airline industry experts said yesterday."

ALAMEDA FREE LIBRARY

USAir officials had tried last-minute damage control, but to no avail. General counsel James T. Lloyd sent a letter to the *Times* two days before the story was published that read in part: "It is possible to look through the tens of thousands of reports that accumulate over time and build a picture that distorts the fundamental truths." His argument got a brief mention in the story, but it was overshadowed by a mountain of evidence that said the airline had a safety problem. When USAir officials saw the story, they felt they were victims of a hatchet job.

Rick Weintraub, a former *Washington Post* reporter who had just been hired to be the airline's chief spokesman, quickly put together a fact sheet that criticized the *Times* for making sweeping allegations that were incorrect or misleading. It said the anecdotes about nine planes leaving without sufficient fuel were correct but that most of them returned before taking off. It disputed the *Times*'s claim that pilot mistakes were a common thread in three of the five crashes and another claim that said the airline was in such dire straits that it was losing $2 million per day. The fact sheet pointed out that the *Times* buried more-favorable comments about USAir toward the end of the story.

The *Times* story was correct in reporting that USAir pilots were having difficulty adhering to company procedures. That was the whole point of Operation Restore Confidence, which had begun before the Hopewell crash. The NTSB found notes from August 1994 meetings that showed pilots often did not follow company procedures. When FAA inspectors observed one hundred USAir pilots, only forty-six followed the company's *Pilot Handbook*. Nine days before the *Times* story, USAir 737 flight manager Jim Gibbs held a meeting on the problem of pilots not flying by the book. "We must have failed to either train or enforce the standardization," he said, according to notes of the meeting. "Now we must find a way to correct the problem."

But the *Times* story exaggerated USAir's troubles by giving the story so much space, putting the most damaging evidence on the front page, burying the more balanced comments inside and relying on easy dial-a-quote comments from people who didn't know the intricacies of aviation. (Ralph Nader: "The problem was that these mergers came with a price. They diverted management attention and took a lot of revenue that could have been spent on safety.") USAir was correct in saying that it was easy for reporters to pluck incidents from the FAA's databases to distort an airline's safety record. Indeed, it is difficult to measure the vague concept of safety. The FAA databases are designed so the regulators can spot problems before they cause a crash—not to see which airline is most likely to kill you. Accidents are so rare that it's usually unfair to use them to draw conclusions about a single airline. The *Times* made that point deep in its story, but it also used a chart on the front page that showed the likelihood of dying was more than three times higher on USAir than on the other carriers. And while the lengthy *Times* story was somewhat balanced, the abbreviated versions used by other newspapers around the country were more one-sided. The long, detailed *Times* story got boiled down to a simple message: USAir was unsafe.

The impact was swift and powerful. The airline's bookings plummeted again. Airline analysts said they were increasingly worried about the possibility of a bankruptcy filing. Employee morale sank to new depths, as ticket agents suffered through wisecracks from customers. Three days after the story was published, Schofield sent a bulletin-board message that tried to give everyone a boost. "I salute you for your patience and professionalism in handling these pointed conversations calmly and with confidence," he wrote. He quoted aviation expert John Nance from a TV appearance saying, "You can take any airline in the country and find examples of things that sound this bad when taken out of context, in isolation, raise them into scrutiny, and scare everybody to death. But when you really look at this airline's heart and soul of operations, they're no less safe than any other major carrier."

Schofield urged employees not to be bitter. "I know that it is painful to see and hear negative and distorted media coverage. Our best response, however, is to prove critics wrong by action, not words. We know we're a safe and reliable airline. We must continue to demonstrate that to our customers in every action we take and every contact we have with them."

Five days later, USAir launched a major public relations campaign to try to rescue its reputation. It appointed retired U.S. Air Force general Robert C. Oaks as vice president for corporate safety and regulatory compliance. USAir also said it was hiring PRC Aviation of Tucson, Arizona, "a respected and experienced aviation consulting company," to conduct a thirty-day audit of the airline's safety policies and procedures. Schofield said the auditors "can go anywhere, ask anyone anything, can look at any records, manuals, bulletins, letters or messages they think are germane to safety at USAir. There are no limits."

The airline then bought full-page ads in major newspapers to publish messages from the company's employees and unions. The ALPA chairman's letter said safety was the pilots' first priority. The head of the flight attendants union said USAir was totally dedicated to "operational integrity." Two mechanics said they would not hesitate to ground any plane if it was unsafe. A customer service supervisor said passengers' well-being was foremost in the minds of USAir employees. Schofield concluded the series with a message that said safety "is the foundation for all we do."

John Cox sat down at the conference table in the NTSB listening room and put on a set of headphones. He and USAir pilot Ed Bular had come to hear the cockpit tape from Flight 427 to see if they could identify the mysterious thumps that had baffled the investigators for months. Cox had heard tapes from other crashes before, and he could never erase them from his memory— the routine of the cockpit quickly deteriorating into shouts, screams, and death. But listening to the tapes was a necessary part of an accident investigation and, as a pilot with eight thousand hours in 737s, he might recognize something that others did not.

He and Bular sat across from each other at the small table. Al Reitan, an NTSB sound technician, sat at the head of the table so he could control the tape player. They began by listening to all the cockpit sounds for the final ten to fifteen minutes of the flight. That was a necessary step for Cox, to listen to the entire tape so he could get over the drama of the event, the fact that he was listening to two fellow pilots scream and die. He was amazed by the sound of the plane being buffeted, a violent shaking caused by insufficient air crossing the wings to keep the plane flying.

"The buffeting sounds like a goddamn freight train," Cox said.

Reitan then isolated each of the four channels on the tape—Germano's microphone, Emmett's, the jump seat microphone, and the mike in the ceiling above the pilots' heads. Both pilots sounded cool and confident, Cox thought. There was good rapport between them, and they were flying by the book. Emmett said very little in the final thirty seconds. His most expressive comment was a worried "Ohhh shiiiiit." But Cox thought that was understandable because he was the flying pilot and was trying to figure out what was happening to his airplane. Germano was more expressive, but he never said what he believed was happening. His most telling comment was when the plane stalled, when he asked, "What the hell is this?"

Reitan used filters to block out the extraneous noise so the pilots could focus better on the voices. He used a computer to trap snippets of sounds and play them repeatedly. He trapped the thump and played it back again and again. *Sheez Zuh Thump Clickety-click. Sheez Zuh Thump Clickety-click. Sheez Zuh Thump Clickety-click.* Over and over the pilots listened to it, like an old record album with a scratch on it. But after hearing it dozens of times, they could not identify the sound. "Beats the hell out of me," Cox told Bular. He thought it sounded like a briefcase falling over on a carpeted floor. But cockpits were not carpeted, and there was not much space for a briefcase to fall over.

Several hours later when Cox walked out of the room, he was convinced that the pilots had been caught by surprise by something they hadn't encountered before. He believed the tape proved the pilots had not done anything wrong. It offered no evidence that one of them stomped on the left rudder pedal and held it down. To the contrary, it showed the airplane was doing something the pilots did not expect.

Cox said later, "There is a gremlin in that airplane."

McGrew, Boeing's chief engineer for the 737, listened to the same tape and came to the opposite conclusion. He had to fight for months to get the NTSB to allow him to listen to it, because he was not officially part of the investigation. When he finally heard it, he used the same cool approach that he used for everything else. He concentrated on gathering data, hearing every sound he could, and didn't think much about the screaming.

McGrew came away convinced that the pilots were startled by something and then overreacted. He heard it in the tone of their voices when they said "Sheez!" and "Zuh!" They had no idea what hit them. McGrew heard nothing to indicate that the pilots believed the plane was malfunctioning. There was no mention of the rudder pedal or anything else not working properly. He noticed they never communicated about the fact that they were twisting out of the sky. When Germano shouted, "Pull! Pull!" McGrew knew it was the wrong thing to do. They should have pushed the stick forward to gain speed.

The tape made it all clear to him: the pilots got startled, stomped on the pedal by mistake, and then pulled back on the stick, stalling the airplane.

The crash was clearly caused by pilot error. He wanted others to listen to the tape because he was convinced they would have the same reaction he did.

Months after hearing the tape, McGrew awoke in the middle of the night, haunted by the sounds he had heard. He wondered what the passengers felt as the plane was spinning toward the earth. What do you think about when you know you're about to die?

The different interpretations of the tape showed how the rivalry between ALPA and Boeing was increasing. As the investigation wound into its fifth month with no conclusive evidence as to why the rudder had moved, the two groups were beginning to disagree more often. In the absence of solid facts, both sides retreated to positions that protected their turf. It was as if Boeing and the pilots union had a default setting, like computer software. Until they got proof to the contrary, they didn't budge.

The two sides remained cordial, but the rivalry was apparent. Cox enjoyed taking friendly jabs at the Boeing team. He tapped McGrew on the shoulder at their first meeting and said, "I know what did it." Cox held up a copy of the *Weekly World News,* a supermarket tabloid that had this headline: USAIR FLIGHT 427 COLLIDED WITH A UFO! The article reported: "Federal investigators are looking into the possibility that the crash of USAir Flight 427 was caused by a collision with a UFO—a possibility supported by the discovery of a passenger's hastily scribbled note that says, 'Massive, glowing, as big as a house. Oh my God! It's going to hit!'"

McGrew got a chuckle out of it, but he and the other Boeing investigators were wary about Cox and the rest of the ALPA guys. They felt that ALPA was overprotective of the pilot brotherhood and would go to great lengths to protect a fallen pilot, even when he was to blame.

Brett decided to go back to Pittsburgh for the January 1995 public hearings on the crash, where the safety board would lay out the evidence it had amassed. He chose to fly because it simply wasn't practical to drive from San Diego, where he had spent several weeks staying with his brother. But he had been very careful about choosing how to get there. He picked American Airlines because it was the only major carrier that flew to Pittsburgh and did not have any 737s.

As the plane prepared for takeoff, the man sitting beside him noticed Brett was clenching his seat.

"Nervous flier, huh?" the man asked.

"Yeah," Brett said.

"They say your odds of dying in a plane crash are higher than winning the lottery."

Brett couldn't let that go. He reached into his wallet and pulled out the card about the scholarship fund he had started in honor of Joan.

"That plane that crashed in Pittsburgh," Brett said, "my wife was on it."

"Oh my God," the man said.

Then the man told Brett about his own personal tragedy, that his daughter had been molested by his ex-wife's new husband. Brett had heard lots of those sad stories since Joan died. When people found out about his tragedy, they wanted to share their own horrible tales, reassuring him that he wasn't alone in suffering. A *Newsweek* photographer told Brett that his brother took a drug overdose. A cab driver said his parents died when he was twelve. A woman who handled Joan's pension said her brother and his wife were in an accident with a drunk driver, which had killed their child and put them both in comas. Tragedies everywhere. Until this happened, Brett had no idea that life could be so miserable for so many.

The man beside him said he was a born-again Christian and that he had decided to forgive the guy who had molested his daughter. "I've realized that if I want to be forgiven by God, I've got to forgive the people who harm me." That really struck a chord with Brett. He had never really bought all the hype about being born again, but the guy was right. Brett realized that he couldn't stay angry forever.

The next day he took a cab from his downtown hotel out to Hopewell to visit the crash site again. He brought a video camera so he could record the scene for Joan's family. He had talked with them often in the five months since the crash. He hired an artist to do an oil painting of Joan from a photograph and gave it to her parents for Christmas.

He trudged through the snow, narrating while he described the site. He took a wide shot of the spot where the nose of the plane struck the road. "You can see it's not very big at all," he said. "It's not much bigger than a family room, really."

He walked to the plastic wreath, which sat on a big easel with American flags on one side. "This is that wreath I was telling you guys about, where Joan's rose blew off just as the guy from the Salvation Army was walking up to it. There are still a lot of them on there. They are glued on there pretty good. It's still kind of an amazing thing to me."

He then visited the Sewickley Cemetery, where USAir had built a memorial that listed all 132 victims. He zoomed in on Joan's name and then read the inscription on the headstone very quickly, as if he was in a hurry to leave: "Strengthen our course with every prayer, let Heaven's breezes speed us there, and grant us mercy evermore as we sail to Heaven's shore."

13

PILOT ERROR

Pittsburgh was brutally cold the week of the public hearing. Several inches of snow covered the lawn outside the Pittsburgh Hilton and Towers, and the roads were filled with an ugly gray slush. The hearing was conducted in the hotel's ballroom, a cavernous space with nineteen-foot ceilings that was set up like a giant courtroom. Haueter and his investigators sat at raised tables on one side, as if they were the prosecutors in the case. The parties—Boeing, the pilots union, the FAA—were positioned just below Haueter, huddled at tables as though they were the defendants. The meeting was run by NTSB chairman Jim Hall, who presided like a judge. In the audience were several hundred spectators, including about one hundred relatives of the victims. Some of them clutched photographs of Flight 427's passengers. Others had photos pinned to their jackets and sweaters.

The NTSB called it a public hearing, but that was a misnomer. Many families had the impression that they would be able to stand up at a microphone and ask questions, just as they would at a city council meeting. But at an NTSB hearing, the public was to be seen and not heard. This meeting was the agency's chance to show its work and explain the evidence. There would be no time for questions from the crowd.

There would be no breakthroughs, either. Haueter and his investigators had spent the past five months talking with the people who would testify, so it was unlikely that the NTSB would learn anything new. If a break in the investigation ever occurred, it would be in a lab somewhere, not in a big ballroom in Pittsburgh.

On the first day of the hearing, the board released the docket, a foot-high stack of reports, memos, letters, charts, and graphs. Much of it was technical gibberish—details about dual-concentric servo valves, rudder blowdown, and chip-shear strength. There were pages and pages of data from the flight recorder and photocopies of federal aviation regulations. Nowhere in the big package was there any guidance to understanding what it all meant—no underlined words, no comments in the margins, no Post-it notes. It looked like Haueter had emptied his file cabinet and ordered somebody to make copies. But it was consistent with the cryptic way the NTSB operated. It released hundreds of thousands of facts and allowed the public to decide which ones were important.

Haueter began the hearing by reading an account of the five-month investigation: "On September 8, 1994, at about 7:03 Eastern Daylight Time, USAir Flight 427, a Boeing 737-300, registration N513AU, crashed while descending to land at the Pittsburgh International Airport." He played a video animation of the moments right before the crash. The video showed a plain white 737 (USAir had balked at putting the airline's logo on the animated plane) that bobbled a bit, rolled smoothly to the left, and then plunged nose down. The video ended while the plane was still at an altitude of 4,000 feet. (A Boeing video played later in the day was more dramatic. It showed the view Emmett and Germano would have seen, with the ground spinning closer and closer until impact.)

Haueter explained the basics of the investigation—the recovery of the wreckage, the reconstruction in the hangar, the backdrive work in the simulator, and the multitude of tests on the rudder system. He didn't say they were stumped, but he made it clear that they had not found the answer. "Mr. Chairman, at this time, I am not aware that any party to the investigation or any other persons or organizations have raised avenues of investigation that we have not pursued fully, or are currently examining."

One by one, Haueter, Phillips, and other NTSB officials asked the witnesses to tell what the NTSB had found. Haueter and Phillips knew the answers to most questions before they asked them; the point of the big production was not to uncover new facts but to let the public hear the old ones. It was peculiar—NTSB investigators asking outsiders to describe what NTSB investigators had done—but that was how the agency worked. In the safety board's just-the-facts-ma'am culture, it was preferable to let outsiders take the spotlight. That approach allowed the investigators to be impartial and kept them from speculating publicly about the cause.

Bill Jackson, the pilot who had ridden in the cockpit on Ship 513's previous leg, testified how his knee was pressed against the microphone button, which would explain the gurgling sound the passenger heard in first class. An FBI agent testified about the examination of wreckage and how the agency had ruled out the possibility of a bomb. A parade of Boeing engineers explained the NTSB's efforts to jam the rudder PCU, the simulations in M-Cab, and the tests on hydraulic fluid. All showed that the rudder and the 737 had performed properly.

McGrew had never testified at an NTSB hearing, but when he walked into the ballroom and saw the families sitting in a special section he realized that the purpose of the hearing was not to advance the investigation. This hearing was just a big show, a way for the NTSB to get publicity. After answering general questions about his education and his job at Boeing, he used the remaining questions to convey his main points—that the plane had passed every test and that Boeing was committed to safety.

He said his boss had told him "to go out and find the cause and, if it's anything to do with the airplane, fix it." He spoke proudly of his plane and its safety record, like he was boasting about his kid's SAT scores. He wasn't basing this testimony on some emotional tie he had to the plane, of course. It was all based on hard data. And the data showed his plane was incredibly safe. More than 2,600 737s were now in use in ninety-five countries, he said. The 737 had the best reliability of any airliner and—this was the data talking, mind you—the rate of 737 hull losses was extremely low.

Unfortunately for Boeing, McGrew did not come across well on the witness stand. His cool reserve about the plane's record made him appear smug, and much of his testimony sounded rehearsed.

Toward the end, Phillips tossed him a softball question, asking if there was anything else he wanted to say. McGrew seized the opportunity, pointing out that the PCU from the USAir plane would not reverse, that the fluid was not significantly contaminated, and that there was no evidence of a jam. "That leads us, based on that data, to think that the rudder was doing what it was asked to be doing."

In other words, the pilots did it.

The most unusual event of the week came during the lunch break on the second day, when about twenty family members held a press conference in a tiny meeting room down the hall from the ballroom. They took turns stepping up to the microphones to complain about the shoddy treatment they had received from USAir. "We believe the system for notification of next of kin is deeply flawed," said Marita Brunner, whose brother-in-law Jeffrey Gingerich was killed in the crash. "It increases the anguish for the families."

Joanne Shortley said she started calling USAir the minute that she suspected her husband, Stephen, was on the plane, but all she could get was a busy signal. When she finally got through, a USAir employee took her phone number and said the airline would call back. Her brother called the airline and was told that Stephen was not on the flight. Joanne's children cheered. But hours went by and Stephen did not come home. Finally, the airline called back at 2:45 A.M. to say that he was on the plane. Like Brett, Joanne thought that her USAir coordinator was poorly trained and unprepared for the family's grief. "She was not equipped in bereavement," Joanne said at the news conference. "She was a saleswoman."

Judy Lindstrom, whose husband, Gerald, was killed in the crash, complained that USAir had blocked her efforts to get a list of other families. "We had a great need to see each other and be with each other. I was told this was not something the airline would disseminate. We found out later that many of us had made the same request. We had to scramble and scratch to get together." The families said they were forming the Flight 427 Air Disaster Support League to urge the government to appoint a family advocate who would help families after a crash. "We are demanding that this process be taken away from the airline," Marita Brunner said.

Brett agreed with the group's complaints, but he didn't attend the news conference. He had kept his distance from the group. He wasn't much of a joiner, and he did not want to get too wrapped up in the crash. It seemed as if some people in the group wanted to make the crash the centerpiece of their lives. He didn't. He knew that he had to move on. He spent much of the week reading the docket in his room and exercising in the fitness center.

Reporters left the news conference and cornered Deborah Thompson, the USAir director of community affairs, on the mezzanine outside the ballroom. Was it true, they asked, that USAir had done a poor job?

Thompson said the process of identifying the passengers had gone slowly, but that the airline wanted to be sure the list was accurate. "You don't want to give wrong information," she said. She acknowledged that the family coordinators had not been trained and said that the airline planned to start training people so they would be better prepared in case of a future crash. "We want to do a better job. But that's not to say I think we've done a bad job."

Jim Hall, the NTSB's new chairman, had just replaced the retiring Carl Vogt. A short man with sandy hair and a mild Tennessee drawl, Hall was a Vietnam veteran, a longtime Democratic activist, and a friend of fellow Tennessean Al Gore Jr. When he was nominated for the job in 1993, *Washington Post* columnist Al Kamen called Hall "a politically connected white male Democrat whose only transportation experience apparently is a driver's license."

Hall appeared to be a lightweight because he didn't talk like an engineer. He used folksy Tennessee phrases and often sounded like Andy Griffith on *Matlock*. While other people were talking about dual concentric servo valves, Hall would be recounting something his mother had taught him. It didn't help that he had a dog named Trixie in his office.

A friendly brown-and-white Welsh Corgi, Trixie belonged to Hall's special assistant, Jamie Finch. Hall was an animal lover, but his three dogs and four cats were back home with his family in Tennessee. So he was glad to have Trixie around, even though she occasionally pooped on his carpet. He would throw tennis balls for her and reach down and pat her during meetings. At Christmas, he and Trixie walked through the office wearing matching Santa hats. She even had her own NTSB badge that said her title was "Safety Dog."

The truth was that Hall was not a lightweight at all. He was a savvy political operator who knew how Washington worked. He was a loyal soldier in the Clinton-Gore crusade to make government more responsive to taxpayers. He was well connected at the White House and in Congress, so he had the clout to get the money and staff that the NTSB needed. But he also had an outsider's perspective and would fuss at the NTSB engineers to make sure that they explained their findings in everyday English. In the NTSB's ongoing fights with the FAA, he used his political skills behind the scenes and then, if that didn't work, he would offer a few choice comments to the media.

Before Hall arrived, the safety board had not paid much attention to victims' families. Investigators would politely answer questions, but they thought their job was to find out why planes crashed, not to provide grief counseling. When Hall heard that the families were unhappy and had formed a support group, however, he told the NTSB staff to help them. In his view, helping the families was exactly what the Clinton-Gore approach called for. He asked the NTSB staff to set aside reserved seats in the ballroom for the families and arranged a meeting with them one night after the testimony. He had Haueter come along to answer questions.

About a hundred family members came to the meeting room on the mezzanine level. Hall opened the session by explaining the NTSB's role and the purpose of the hearing. Haueter updated the families on the investigation and went through the plans for the remainder of the hearing. Then it was time for questions. The 737 sure seemed to have lots of rudder problems, someone said. Why not ground it?

That wasn't up to the NTSB, Haueter said. His agency would only make recommendations to the FAA, and it needed solid evidence before it made such a serious request. So far, he hadn't found the evidence.

Why have Boeing people up there? someone else asked. "They're just going to lie to you."

Haueter and Hall explained the party system. "These people know the plane's systems best," Haueter said.

Someone else was amazed about the lack of measurements on the flight data recorder. "Why aren't you doing anything about that?"

"Look, you're preaching to the choir. The safety board has been saying that for years," replied Haueter. Hall vowed to bring up the issue with the FAA.

The families had a litany of complaints about USAir—the long delays in confirming who was on the plane, the poorly trained employees, and the airline's refusal to share the names of other families.

Hall listened and came away convinced that the families had been mistreated. The things they wanted were reasonable. USAir didn't have to pamper them, but the company should have shown a little common decency. Hall would have preferred that the airlines do it on their own, but they had shown they did not care enough to do it right. It was time for the government to get involved.

The hearing revealed growing tension between the NTSB and Boeing. The company had fought some of the NTSB's requests for data, had stamped "PROPRIETARY" on several items and said the board could not release them to the public because of Boeing's confidentiality agreements with the airlines. Days before the hearing, the NTSB learned about several 737 incidents that Boeing did not include on a list that Phillips had requested.

Haueter was surprised that the company was so disorganized. He didn't think Boeing was deliberately hiding anything, but he was disappointed that it could not keep track of important data. Chairman Hall was furious. He reminded a Boeing official of the request for *all* incidents and asked, "Is that simple enough?"

Another Boeing official acknowledged that the company should have included a rudder incident involving an Air France 737 on the list, but the report did not sound serious to Boeing employees when they first heard about it. When the testimony concluded, Hall told McGrew he was concerned that the company's list of 737 incidents was incomplete. "When we end up in a situation, Mr. McGrew, just to be straight with you, that we request information and then another party sends us information that is pertinent that we didn't get from you, it causes concern." Hall said he knew that things fell through the cracks, but he told McGrew to "go back and examine every crack so we don't have any question that there's been any incident with this rudder or any of these systems that might assist us." Hall then apologized for being so harsh. "I'm from Tennessee and I don't know to express myself any more than just that way."

McGrew was angry. He felt like he had been used so Hall could get his name in the press. The NTSB staff knew all the details about the Air France rudder kick. Either they didn't tell Hall or Hall conveniently ignored it so he could land a few punches on Boeing. McGrew did not like the fact that politicians were getting involved in plane crash investigations. The NTSB was becoming a political beast, under enormous pressure to come up with an answer. And McGrew was not sure they would come up with the correct answer.

As the hearing concluded, reporters gathered around Haueter and Hall to get their reactions to the week. Haueter had consistently told the press that he was confident he could solve the mystery. A reporter asked how he could be so confident with so little good evidence.

"It's based on experience and the data available," Haueter said. "I've seen two other accidents I've worked on where I had much less data than this, much less help than this, and we determined probable cause very definitively. I can't say I identified any new alleys [to investigate] this week . . . but there are a lot of avenues available that have to be fully explored."

As McGrew returned to his office in Renton, he was more confident than ever that nothing was wrong with his airplane. It bothered him that critics were saying Boeing might be covering up a hidden flaw in the plane. McGrew said he felt no pressure from lawyers or anyone else to protect the plane. If the 737 had a flaw, Boeing wanted to find it.

McGrew approached that mission with vigor. He spent hours reading reports on the crash and led daily meetings of Boeing engineers. Ninety-five Boeing employees had worked 42,000 hours on the investigation, with twenty-three of them dedicated to it full-time. It had cost Boeing $1.5 million. McGrew had become obsessed with the crash. Occasionally, he awoke at

3 A.M. with questions about a theory, crawled out of bed, put on his bathrobe, and went downstairs to his home computer. He would pull up a spreadsheet of 427's flight data and try adjusting ratios and parameters to see what effect they would have. Then he would go into the office that morning and ask one of his engineers to try a new computer run.

He kept a list of theories titled "Items Under Consideration." It had eighty-five possibilities, including everything from bird strikes to thrust reversers. All but eleven had been ruled out. Many of those eleven were long-shot theories that were likely to be disproved by future tests. The investigation had largely become a tug-of-war between two theories—a rudder system malfunction and a mistake by the pilots.

From the start, McGrew had been open to the possibility that something was wrong with the 737. That was the classic engineer's approach, to consider every possible failure mode of your creation. That's why redundancy is such an important concept in aircraft design. If something fails, there is at least one backup, often two. The rudder system had lots of redundancy. Every lever inside the power control unit had a second lever that moved in concert, in case one should break. In the valve itself there were two slides in case one should jam. It was powered by two hydraulic systems in case one should fail. And there was a standby actuator with its own valve in case the main PCU stopped working.

But McGrew's plane had come up clean. Test after test failed to find anything wrong with the 737 rudder system. The PCU would not reverse, the valve didn't leak, and there were no marks to indicate that it had jammed. The hydraulic fluid from the USAir plane had been dirtier than it was supposed to be, but tests found that made no difference. You could make the fluid as thick as Dijon mustard and the valve could still do the job. The bottom line, as summed up by Phillips in his report, was that the PCU was "capable of performing its intended functions." Besides, the worldwide fleet of 737s had an extraordinary twenty-seven-year record, with 65 million flight hours and an extremely low crash rate. If there was a flaw in the rudder system, it would have shown up years earlier. McGrew felt it was time to take a closer look at the human part of the equation—the pilots.

The NTSB had spent weeks studying and testing the PCU, but McGrew felt the investigation had barely looked at Emmett and Germano. What kind of training had they had? What did previous incidents tell about how they responded to wake turbulence? The NTSB had not explored these areas very much.

McGrew kept using the word "startled." There was plenty of evidence that the pilots had been surprised by something—maybe a bird, the wake turbulence, or another plane—at the moment that Germano said "Sheeez!" The surprise could have caused Emmett to jerk the wheel too far to the right. When the plane quickly rolled in that direction, Emmett might have tried to stop the roll by slamming his foot on the left pedal and continuing to press on it without realizing what he was doing. There also was no question that the pilots had made a huge blunder by pulling back on the control column. That stalled the airplane and gave them virtually no chance to recover. Yet that mistake got virtually no attention from the safety board.

On February 15, 1995, Boeing went on the offensive and faxed Haueter a seven-page letter that said the NTSB should take a closer look at the pilots. It cited a British report that attributed many military accidents to "overarousal," when pilots were so surprised by an alarming situation that they could not recover. The letter said a *perceived* emergency was all it took to startle the pilots (a subtle reference to the moment when Flight 427 was jostled by the wake turbulence). The Boeing letter also cited an FAA study that said a pilot might take up to ten seconds to respond after being startled. "The NTSB should explore, from a review of the literature and all available databases and records, whether the Flight 427 flight crew could have responded to the unexpected and startling encounter with significant wake turbulence by (1) making an inadvertent application of left rudder, or (2) having an accidental or cognitive failure that led to an application of left rudder," the letter said.

Cognitive failure. That was a delicate way of saying the pilots screwed up.

The letter said there were numerous examples of accidents caused by pilots stomping on the wrong pedal. The 1985 crash of a Midwest Express DC-9 was caused by pilots who responded to an engine failure by pushing the rudder the wrong direction. A 1992 Air National Guard crash of a C-130 was caused by the same mistake. The letter asked Haueter to explore Emmett's and Germano's backgrounds to see if they had been trained to use the rudder or had flown other planes in which they would have used it more heavily. The letter said Haueter also should look into the fact that the pilots pulled back on the control column.

Haueter had mixed feelings about whether they would have stomped on the rudder pedal. His ride in M-Cab had made him doubt that two seasoned pilots would have made such an obvious error. But like the Boeing engineers, he was concerned about pilots becoming too complacent in modern cockpits. Unlike the "stick-and-rudder" pilots of the old days, modern crews had become too much like computer programmers. They relied heavily on the autopilot and flight-management computers, which could practically fly an entire trip. So Haueter agreed to delve more deeply into Emmett's and Germano's training. Besides, the investigation had sunk into a lull and he had been looking for something to get everybody thinking again. The letter was just the jolt he needed. He faxed a copy to the pilots union.

It didn't take long to get the predictable response. ALPA went ballistic. Suggesting that pilots screwed up was akin to shouting a racial slur at them. Herb LeGrow, ALPA's coordinator for the crash investigation, thought that Boeing was looking for a scapegoat. "We don't want to see the reputations of the pilots compromised because [the safety board] can't find an answer to what caused the accident," he said in an interview in his Clearwater, Florida, home. LeGrow, a USAir 767 pilot who had worked on more than a dozen accident investigations, was also worried that Boeing would use its Washington clout to pressure the NTSB into blaming the pilots. He said ALPA might be small, but it was not afraid to go head to head with Boeing. "It's David and Goliath at this point. If it gets down and dirty, I'm willing to fight. We'll sharpen up our slingshots and fight them."

LeGrow and Cox fired off a letter to the safety board saying that Boeing was trying to raise doubts about the pilots and divert the investigation. The letter countered each point that Boeing had made. It said the cockpit tape showed that the pilots were not overreacting to the wake turbulence but were "struggling in an attempt to gain control of an uncontrollable aircraft." The letter said it was unfair to make a comparison with military crews because airline pilots fly considerably more hours. In ALPA's view, there was no need for a further exploration of Emmett's and Germano's backgrounds because the investigation so far "revealed that these individuals were fully qualified and had exemplary records."

LeGrow and Cox said there was no connection between the USAir crash and the pilot-error accidents cited by Boeing. The Midwest Express and C-130 crashes were engine failures at low altitude. "In USAir 427, there is no evidence that the flight crew applied an inappropriate flight control input. In fact, there is a significant amount of evidence which could lead to the conclusion that the aircraft experienced a mechanical malfunction." The ALPA letter also defended the pilots for pulling back on the control column: "It should be noted that at the onset of the event, traffic beneath USAir 427 was a real issue. At that point, maintenance of altitude was, in fact, critical. No action by the crew could have stopped the roll. . . . By the time control column position became an issue, ground impact was inevitable."

LeGrow concluded the letter by saying that the pilots "fought for the lives of their passengers" and suggested that Boeing's letter was written by its lawyers: "Everyone recognizes the manufacturer's product liability problem. The issues of civil litigation should not be allowed to infiltrate an NTSB investigation. The traveling public deserves the answers to what truly caused this accident."

McGrew was not surprised that ALPA had a near meltdown over the Boeing letter. That was typical of the union, he thought. It always seemed to want to protect the brotherhood, even when the facts might suggest otherwise. It seemed as if ALPA wanted to perpetuate a myth that every pilot was perfect and never made mistakes.

The roots of ALPA's defensiveness about pilot error dated back to the late 1920s and early 1930s, when airline managers pushed pilots to fly long hours and take risks in dangerous weather. Pilots who tried to be safe often got fired. At least twelve of ALPA's twenty-four founders were killed in accidents. But when a plane crashed, "pilot error" seemed to be the government's automatic response. Rarely, if ever, did the airline get blamed, even if it had ordered the pilot to fly into a snowstorm. It was easy to blame dead pilots because they could not defend themselves. Ever since then, one of ALPA's fundamental principles has been to clear the names of dead pilots.

When an ALPA member got blamed for a crash, it was a black eye for the entire organization. The union also had a stake because it acted like a law firm, defending pilots who were accused of violating federal air regulations. That was part of the reason for such high union dues (about $4,000 per year for a 737 captain). If pilots got into trouble, the union provided a lawyer and a technical representative (a pilot such as Cox) who defended the accused during the hearings and appeals. The union was good at representing its members, often convincing airlines and the FAA that they should reduce or drop charges, but that effectiveness made people at the NTSB, the FAA, and Boeing skeptical when ALPA said it would be unbiased in a crash investigation. How could the union defend its members and be unbiased?

It was rare that you ever saw Cox sweat, let alone make a mistake. He had tremendous self-confidence, which is why you felt so good with him in the captain's seat. But on one flight in 1973, he got so frightened that he was afraid everyone on the plane was going to die.

He was nineteen, home from college for Christmas break and flying occasional trips as copilot on a Cessna 421, a two-engine propeller plane. On one flight he was assigned to be copilot to take several businessmen from Birmingham, Alabama, to Erie, Pennsylvania. The plane was heavily loaded with people and luggage, so Cox and the captain planned to stop in Pittsburgh and refuel. The plane picked up ice as it descended through the clouds to Pittsburgh, but they were not worried. Rubber boots on the leading edge of the wings inflated to crack the ice so it would fall off, and the propellers were heated so ice wouldn't build up on the blades.

They landed in Pittsburgh without difficulty and spent about thirty minutes getting refueled. On the ground, they made a cursory check of the weather, but only for Erie. They had gotten through the clouds around Pittsburgh without difficulty and figured the weather there was no big deal.

They took off and started to climb back through the same clouds, which topped off at 5,800 feet. As they climbed, they realized that the small Cessna was taking on a tremendous amount of ice. It got so thick on the wings that the boots stopped working. Then they heard a horrible sound like a machine gun. *Bam-bam-bam-bam-bam-bam.* It was the sound of the propellers hurling ice at the fuselage. The ice was building up so fast that the plane was getting dangerously heavy. They had the engines at full power, but the plane could barely climb.

Cox looked out the window and saw ice growing on the wingtip fuel tanks. It was forming a menacing-looking icicle a foot and a half long and growing by the minute. Another blob of ice had developed on the spinner, the hub at the center of the propeller blades, and it too was growing. The windshield was so covered with ice that it was opaque, and the pilots had diverted all the heat in the plane to the vents on the windshield. That made the cockpit extremely hot and left the passengers freezing in the back, but it was having little effect on the windshield. If Cox or the captain wanted to look outside, he had to squint through a tiny strip right above the vents where the ice had melted. Cox was sweating a lot—a combination of heat and fear.

The ice was so thick and heavy that the plane lost its ability to climb, even at maximum power. Cox and the captain discussed what they should do. They didn't want to return to Pittsburgh because they would have to go back through the nasty clouds. So they decided to continue to Erie, where the weather was slightly better. At that point, Cox wondered if they would survive. He wasn't thinking about dying, exactly. He was worrying about what would happen if they had to set the plane down in trees. It would rip the aircraft apart. At that point, dying would be a foregone conclusion.

As the captain flew the plane, Cox radioed to air traffic controllers, telling them the plane had a severe ice problem and that they were descending to the MOCA—the minimum obstacle-clearing altitude. It was the lowest they could fly and not worry about crashing into mountains and radio towers. Now that they were descending, the ice appeared to stop growing. It hadn't shrunk, but at least it wasn't getting worse. Cox was sweating so much now that he had soaked through two shirts. He normally wore his shirts buttoned all the way up and his ties tight around his neck, but now he had the collar unbuttoned and his tie loosened. His heart was racing.

When he flipped the lever to lower the landing gear, he heard the strain of the electric motors trying to open the gear doors. *Grind, grind, grind.* Then he realized what was happening: The doors had been frozen shut by the thick layer of ice on the belly of the plane. The pilots might never get the gear down. They would have to make a belly landing.

Grind, grind, grind. It sounded like the motors were ready to give up. *Crack!* The right gear door burst through the ice and opened. The gear went down. The plane wobbled a bit from the sudden drag on the right side of the plane.

Crack! The left door opened and down went the gear. They could land now . . . if they could only see the runway.

The captain had used the instrument landing system to line up and descend toward the runway. The ILS was designed for moments precisely like this one, so pilots could approach a runway without seeing it until the last minute. They squinted through the slit in the windshield, searching for the runway lights. As they finally saw the lights and prepared to land, Cox was thinking, *If we keep the plane lined up, we should be okay.*

Thirty feet above the runway, the plane stalled and plunged nose down toward the pavement. There was so much ice that it changed the aerodynamics of the aircraft and made it stall sooner than expected. The captain pushed the throttle lever forward and jerked back on the control column to bring up the nose.

Bam! The plane hit the runway on all three landing gear simultaneously. Ice broke from the belly and smashed into a million pieces on the runway. They taxied the plane to the terminal and opened the door. As the passengers got off, they teased the captain about the hard landing. "That was one of your worst," one passenger said. Cox and the captain gave each other a knowing look. If the passengers only knew how close they had come to dying!

Thinking back on that flight, Cox could see several mistakes. They were in a hurry and got careless. They didn't check the weather reports as thoroughly as they should have. They should never have flown in such bad conditions. Cox's willingness to admit when pilots made mistakes was one of the reasons he was so well respected at the NTSB. He represented a new generation of ALPA investigators, one that was not so protective of pilots.

Yet, in the case of Emmett and Germano, Cox resisted any suggestion that they were to blame. He would grudgingly concede that it had been a mistake for them to pull back on the stick, but he would quickly add that it was an understandable response. At that point, they were watching the ground loom closer and closer. It was natural that they would want to pull the nose up to survive. Besides, Cox said, there was no procedure for pilots to follow if they had a rudder hardover. Such a situation was not mentioned in the pilots' manuals, and it was not covered in their training. Plus, Emmett and Germano had had virtually no time to diagnose the problem and then decide what to do; it was only eight to ten seconds from the first bump caused by the wake turbulence until the plane was unrecoverable.

Pilot-error accidents have been around as long as there have been pilots. Modern statistics show that pilots are the primary cause in about 70 percent of commercial jet crashes. In a 1992 Anniston, Alabama, crash that Haueter investigated, the captain, a new employee of the airline, became subservient to the first officer. They were both trying to be cool and calm, ignoring the fact that neither of them knew where the airport was. At one point the captain said, "Hopin' no one on here's a pilot," an indication that he hoped passengers wouldn't notice the strange flight path they were taking. Both pilots were lost but would not admit it. They thought they were south of the Anniston airport, but they were actually north of it. The plane crashed in trees seven miles from the airport, killing the captain and two passengers.

Other pilot-error crashes are caused by simple mistakes, often in combination with faulty warning systems. In the Delta 1141 crash that Haueter investigated in Dallas, the pilots forgot to set the flaps, a critical step that gives the wings extra lift for takeoff. That alone should not have caused the crash, since the pilots had to go through a checklist to make sure the flaps were set and the Boeing 727 had a warning system that would sound an alarm if they started a takeoff roll without the flaps extended. But the pilots were rushing during the checklist. When the flight engineer called, "Flaps?" the first officer responded, "Fifteen-fifteen-green light" without actually looking to make sure they were set. The takeoff warning system should have sounded a horn about that mistake, but it had been having intermittent problems and did not sound as the plane started to take off. The plane got twenty feet off the ground, bounced tail-first at the end of the runway, and crashed into an antenna, killing fourteen people.

The numbers on pilot error can be misleading because not many crashes can be attributed to a single factor. Investigators often say a crash is the result of many links in a chain. If any link had been broken, the accident would not have happened. The pilots may make the ultimate mistake, but they may be responding to bad weather, malfunctions in the plane, poor design in the cockpit, lack of training, inaccurate information from air traffic controllers, or a lack of proper rules from the FAA. If that takeoff warning system had been working, the captain of Delta 1141 probably would have stopped the plane safely on the runway. In *Why Airplanes Crash,* authors Clinton Oster, John S. Strong, and C. Kurt Zorn write that pilot error was the *initiating* factor in only 11 percent of airline crashes from 1979 to 1988.

Airplane manufacturers have been remarkably successful at reducing pilot-error accidents with new warning systems that tell pilots when they are in trouble. The devices are like fussy robots in the cockpit that shout at the pilots when there is trouble nearby. All airliners in the United States now have a traffic collision avoidance system, or TCAS, that shows other planes on a screen and yells "TRAFFIC! TRAFFIC!" when one gets dangerously close. (That warning was heard on Flight 427's cockpit tape as the USAir plane began to spiral down, probably because the TCAS saw the Jetstream plane a few miles away.) Big jets also are equipped with a ground-proximity warning system that hollers "TERRAIN! TERRAIN! PULL UP!" when a plane gets too close to the ground or a mountain. Those kinds of accidents—known as controlled flight into terrain, or CFIT—have dropped by 43 percent since the warning systems were mandated.

As chairman of the human-performance group, Malcolm Brenner had to explore Boeing's suspicions about the pilots. The cousin of comedian David Brenner, he was the most eccentric person involved in the investigation. A towering six-foot-three psychologist, Brenner had a quick wit and seemed more like a college professor than a crash detective. He got so excited when he discovered good data that he got chills. When it was exceptionally good data, he felt the hair stand up on the back of his neck. He spoke of humans the same way a zoologist talked about giraffes or elephants. Humans were the species that he studied. Never mind that he was one, too.

Brenner grew up in West Philadelphia, studied psychology at Boston University, earned his master's degree at Stanford University, and received his Ph.D. from the University of Michigan. Before he landed the job with the NTSB, he had worked as a psychologist for NASA, an undercover private detective, a limo driver, and a Santa Claus. He said his biggest contribution to highway safety was when he quit being a limo driver.

In the engineer-dominated world of accident investigation, the human-performance experts were the odd ducks. Other investigators dealt with solid, indisputable evidence—bent fan blades that told whether an engine was spinning or a broken gauge that showed a plane's airspeed when it crashed. But human performance dealt with fuzzier issues—what a human did and why.

In the initial probe, the NTSB had tracked down Emmett's and Germano's friends, coworkers, relatives, and everyone else who had seen the pilots before the crash. They were asked about the pilots' moods, what they ate for breakfast and dinner, even how much they slept. A safety board handbook said such checks were important because "it is often possible to find problems in the individual's background that foretell the problems of the accident."

Brenner found that Germano, forty-five, had been happily married for nineteen years and had two young children. He had been a pilot for the air force, Pilgrim Airlines, and Braniff before joining USAir. He had a clean FAA record, no criminal history, and he rarely drank. His pilot training records were generally positive, with comments such as "very consistent and smooth pilot," "nice job," and "excellent landings." The only negative comment had come years earlier from a check pilot who wrote that he rated Germano in the "lower 10 percent."

Brenner sent questionnaires to the pilots' families and scoured public and private records about them. He found that Germano got seven and a half hours of sleep in a typical night and had a ruptured disk in his back that had been removed six months before the crash. He also had mild allergy problems and suffered from a runny nose and postnasal drip. He received a traffic ticket a year before the crash for failing to give a proper signal. He was said to be cheerful the day before the accident, when he flew three legs and had a layover in Jacksonville. He joined other crew members in singing "Happy Birthday" to one of the flight attendants, and he ordered a turkey croissant from room service shortly before midnight.

Emmett had been married for two years and had no children. He had taken flying lessons as a teenager, became a corporate pilot, and then was hired by Piedmont, which merged with USAir. He had clean driving and flying records and no criminal history. A check pilot who rode with him in a simulator test described him as "well-prepared . . . a sharp guy." Fellow pilots described him as "exceptional . . . the kind of first officer you'd want to fly with." He had responded calmly when a plane had a hydraulic failure one month before the Flight 427 crash. His wife said he was an occasional drinker and usually slept eight hours. A USAir chief pilot told investigators that Emmett was a dependable guy who reminded the chief pilot of his son.

Brenner's amiable personality was a big asset in investigating the pilots because he frequently had to call grieving relatives. He had tracked down wives, husbands, girlfriends, and boyfriends to ask about a dead pilot. But running the Flight 427 human-performance group had not been easy. It had some of the strongest personalities in the investigation, especially Curt Graeber, a human factors expert from Boeing.

Graeber had an in-your-face approach that was the polar opposite of Brenner's lighthearted style. A former army lieutenant colonel who had been assigned to NASA and the space shuttle *Challenger* investigation, Graeber believed the NTSB had done a poor job of looking into the pilots' backgrounds. He thought the safety board had asked shallow, easy questions of the pilots' coworkers and ended up with a predictable profile of above-average guys who were loved by everyone. Graeber thought the NTSB should have asked tougher, more pointed questions that would go deeper into the pilots' training records and how they performed in the cockpit.

Graeber viewed Brenner as a weak manager who let the group's meetings drift into pointless discussions. Instead of staying focused on the pilots, the group would go off on tangents about wake turbulence and servo valves and then someone would say they should stick to the facts and the conversation would abruptly end. Brenner seemed unwilling to take charge and keep the group focused. "Malcolm is supposedly the best human factors guy [the safety board has] got, which is a sad statement," Graeber said.

But Brenner felt he had to give everyone a chance to air his or her opinion. He knew they were at the edge of what could be proved scientifically, but he wanted to consider every possibility. To address Boeing's questions, he enlisted a pathologist to study the broken rudder pedals and try to determine if the pilots were pushing left or right at impact. (The doctor concluded left, but that was later called into question because he was not a metallurgist.) He and other group members tracked down further training records to determine if Emmett and Germano had aerobatic experience (none of any significance) and interviewed two Southwest Airlines 737 pilots to see whether they had used the rudder when they encountered a sudden roll in March. (The pilots did, but one pushed right, the other left.) Brenner's group also reviewed medical records for the USAir pilots to see if they had been treated for any problems that might affect their behavior in the cockpit. (They had not, although Germano had gotten allergy shots to treat his runny nose and postnasal drip.)

When it was all added up, Brenner's team had lots of evidence but nothing close to a conclusion. The pilots had little aerobatic experience, but they should not have needed any special training to recover from wake turbulence. Some clues suggested that they had pushed the pedals, but they could have done that right before impact. Ultimately the team was stymied. On June 6, 1995, Brenner's group approved this statement: "There is no way we can conclude for certain that the crew did or did not put in rudder input."

14

DECEDENT

In the eyes of the Cook County Circuit Court, the value of Joan's life would be determined primarily by coldhearted economics—how long she would have lived, how much she would have earned in her lifetime, and how much she would have spent. To calculate that number, economists would use life expectancy tables, details about Joan's health, estimates of her career earnings, and Brett's account of her household chores.

Brett's lawyers would make an initial settlement proposal to USAir's insurance company, based on prior cases and Joan's earnings. But if the company balked at that number—and that was usually what happened—his law firm would hire an economist to do a more scholarly report based on the interrogatories and depositions about Joan.

In previous crashes, airlines and their insurers had occasionally delved into the private lives of crash victims by hiring private detectives to question neighbors and coworkers of the victim. Airlines called it the search for truth, so they could pay a fair price for a claim. In one famous 1986 case, Delta Air Lines fought for the right to mention in court that a passenger was gay. The airline said the man's sexuality was relevant because of the possibility that he might have gotten AIDS and therefore would have had a shorter life. In other cases, airlines have dredged up information about victims' marital infidelities, drug addictions, and alcohol abuse. But only rarely did airlines hire private detectives. Usually the important details about a victim emerged from interrogatories and depositions.

Awards in plane crash cases averaged $2 million, but they could be much higher for people in high-paying jobs with a long career ahead of them. A planeload of young doctors would be a lot more expensive than a plane full of sixty-year-old grocery-store cashiers. The family of a $300,000-per-year surgeon won a $7.4 million verdict after a 1992 USAir crash. The family of Rodney Culver, a running back with the San Diego Chargers who was killed in the 1996 ValuJet crash, received $28 million.

USAir's chief counsel for the Flight 427 cases was a brash Washington attorney named Mark Dombroff. Although Boeing had its own team of lawyers from a prominent Seattle firm, Dombroff and a Chicago law firm had taken the lead for settling the cases. Boeing and USAir never revealed how they had agreed to divide up the liability, but it was clear that Dombroff and the insurance company Associated Aviation Underwriters were speaking for both USAir and Boeing.

Brett's first legal skirmish with USAir and Boeing had been about the venue for the case. Dombroff removed the case to federal court in Chicago so it would be transferred to federal court in Pittsburgh, which was likely to be more favorable for USAir. But Brett's lawyers managed to get the case back to Chicago because they had shrewdly listed Gerald Fox as one of the defendants. Fox was the USAir maintenance chief at O'Hare who had the gurgling-sound conversation with Germano. That gave the case a toehold in Chicago and helped convince a federal judge to keep it in Cook County Circuit Court.

In the file folder for Brett's lawsuit, Case No. 94 L 12916, Joan was referred to as "DECEDENT." USAir's first step in determining the size of the award was to send Brett an interrogatory, a series of broad questions about Joan's lifestyle, education, career, and income. It would give Dombroff and the insurance company an overview of Joan's life so they could begin settlement discussions.

State complete details concerning DECEDENT's employment history, including military service. Include in your description the title, position or rank of each job held, the name and business address of the person or entity employing DECEDENT at each job, the dates each employment began and ended, the type of work involved in each job, the name of the immediate supervisor for each job, the salary, wages, sources and total yearly compensation (including fringe benefits) received for each job, and the date and reason any employment was terminated.

Brett replied that Joan had been a waitress at J.C.'s Cafe in Iowa City, Iowa, while she attended college, earning minimum wage plus tips; she was assistant manager of a restaurant in Vail, Colorado, at $10 per hour; and then she took the job with Akzo Nobel Chemical, where she earned $46,000 per year plus a bonus. For her reason for leaving Akzo, Brett and his lawyers wrote: "Self-explanatory." There was no need to write "killed in plane crash."

> Fully describe DECEDENT's health for the ten (10) years prior to death. Include in your description any injury, illness, diseases, or condition suffered by DECEDENT and, for each injury, illness, disease or condition suffered, describe how any effect was manifested . . . identify all health care providers of any kind . . . and set forth the reasons for said consultation and/or treatment. Also state whether DECEDENT was ever refused life or disability insurance.

Brett and his lawyer said Joan was in excellent health, that her only significant problem in the past ten years had been a kidney infection. She had regular exams with her doctor and gynecologist and routine visits with her dentist. She had never been refused life or disability insurance.

> Describe DECEDENT's eating, smoking and alcoholic intake habits and state whether DECEDENT, at the time of his/her death or at any time during the ten (10) years prior to that date, took any medication (narcotic or otherwise) or other type of drug, whether DECEDENT ever participated in Alcoholics Anonymous or any similar 12-step or self-help group, the identity of such medication, drugs or groups and the reasons for same.

"Decedent's eating habits were normal," Brett and the lawyers wrote. "Plaintiff's decedent consumed alcohol only socially and did not smoke. Plaintiff's decedent did not take any medications (with exception to an occasional aspirin for an ailment), narcotic or otherwise, on a regular basis in the ten (10) years prior to her death. Plaintiff's decedent did not participate in any self-help group."

Brett knew the lawsuit was the only legitimate way he could get revenge against Boeing and USAir. He thought that the companies knew about the 737's rudder defects and rolled the dice with passengers' lives, believing that the malfunction would not occur again. In his view, they made life-and-death decisions for innocent people and seemed to be accountable to no one. Boeing and USAir seemed to think they were above the law.

In his lowest moments during the months following the crash, he became so angry that he briefly fantasized about committing some act of violence against Boeing, to put the company and its families through the same horror he was going through. Those thoughts of personal vengeance and violence were short-lived, however. He realized that the only legal way to retaliate was to hurt the companies in their balance sheets, to win an award so large it would hurt their bottom lines. He hoped an expensive settlement would pressure Boeing to fix the problems with its airplanes. He figured the company might decide it was cheaper to fix the planes than to become involved in additional lawsuits.

Associated Aviation Underwriters, the airline's insurance company, had begun settling cases in January 1995 for an estimated $2 million each. But Brett was in no hurry. He had received about $800,000 from Joan's insurance policies and workers' compensation coverage. He used some of it to set up a scholarship fund in Joan's name to send a young woman from her hometown to the University of Iowa. He also set up college trust funds for each of Joan's ten nieces and nephews. He felt Joan would want him to do positive things with the money. He said, "I want her life to have a contribution to this world."

A *beep-beep-beep* echoed through the USAir hangar in Pittsburgh as the front-end loader tilted the giant Dumpster and then drove in reverse, emptying it on a plastic tarp. The contents came out as a muddy, solid block, like a brown ice cube popped out of a tray. It was March 6, 1995, still so cold that the fifteen volunteers could see their breath.

The volunteers, from the NTSB, USAir, and families of the victims, had come to recover passengers' belongings from the frozen mess. Hundreds of items—watches, teddy bears, books, wallets, and jewelry—had already been returned to families. But many other items had mistakenly been thrown in the trash bin and left outside during the frigid winter. Leaders of the family group, the Flight 427 Air Disaster Support League, had been complaining for months that they didn't get back all the items they should have. After they saw the big trash bins during a visit to the hangar, they kept pestering USAir and the safety board until they got permission to look inside.

The families were still furious with USAir. Not only were they angry about the airline's performance immediately after the crash, but they were also unhappy with the company's decisions about the crash memorial and a burial service. When the airline held a service for the burial of unidentified human remains, families could see two caskets. But the airline did not mention that thirty-eight other caskets had already been buried. USAir, which had followed the advice of the Pennsylvania Funeral Directors Association, said it was trying to be sensitive by avoiding the display of lots of caskets, but the families said they had been deceived.

The memorial was also a point of controversy. Shortly after the crash, USAir had announced that it might buy the Hopewell site to build a monument. That pleased many family members, who wanted a place that would honor their relatives. Several had launched a separate effort to buy the land, but they put their plans on hold when they heard that USAir was interested. The airline scrapped the idea, however, and instead chose to provide three big tombstones at a cemetery ten miles from the crash site.

The granite monument read: IN LOVING MEMORY OF OUR FAMILY MEMBERS AND FRIENDS INTERRED HERE WHO DIED SEPTEMBER 8, 1994.

Many of the families were unhappy with that inscription. It did not mention the crash or Flight 427. Once again, the airline's efforts to be sensitive had backfired. (The airline then built a granite bench across from the monument with the inscription THIS MEMORIAL IS DEDICATED TO THE PASSENGERS AND CREW ON USAIR FLIGHT 427.)

The families were also upset at USAir for refusing to bring the first batch of personal effects to the Hilton during the January public hearing. The relatives had said it would be more convenient to go through them at the hotel, but the airline had refused. USAir officials were afraid the media-savvy family group would turn the whole thing into a big spectacle, with mangled personal effects displayed for the TV cameras.

With all of the other mistakes USAir had made, it was natural to blame the airline for leaving the items in the Dumpster. Mike Benson, an NTSB spokesperson, tried to direct the blame that way, telling the *Pittsburgh Post-Gazette* that the NTSB presumed there was nothing valuable inside, since USAir employees had looked through the Dumpster. But USAir said the trash bins were the safety board's responsibility. The truth was that no one realized there was anything valuable inside. The NTSB investigators were intent on finding significant parts from the plane. The material in the Dumpster looked like burned trash. Haueter knew there were burned, shredded clothes inside, but he never thought a family member might want them back.

"May I have your attention, please!" Cindy Keegan, the NTSB investigator who headed the structures group, called to the volunteers. "What we're looking for here are any personal articles, whether it be a watch or whatever."

One person saw a muddy piece of clothing in the pile and asked, "Did you say you wanted clothing?"

Yes, she said, they should pick out anything that belonged to the people on the plane.

The volunteers wore white coveralls and gloves. A few wore surgical masks, a remnant from the worries about biohazards. John Kretz and Marita Brunner, the leaders of the family group, stood at the back of the pile. When volunteers brought something, Brunner cleaned off the mud and wrote a description of the item on her clipboard:

Casio multi-function calculator—telephone directory, etc.
Sony Microcassette recorder
Woman's yellow vinyl damask manicure set
Pink hair brush
Speedo eye goggles—swimming
Garage door opener Master Mechanic 750MM K722
Green and white striped Dept. of Energy golf umbrella
Accutron man's watch
Macintosh Users Guide for Macintosh PowerBook 160/180
Man's manicure set/maroon leather case
Golf balls—1 Top Flite Logo Reedsburg Country Club; 2 Titleist;
    3 Blue Max Trumbull Marriott; 1 Maxfli DDH-100
2 hair brushes, 1 Avon
1 pair scissors

A surprising number of items had passengers' names on them. Dennis Dixon, whose wife was killed in the crash, found her shredded briefcase. The volunteers found a luggage tag and business card for David Huxford, a Maryland computer consultant who was heading home after a business trip. They found Kevin Rimmell's Blockbuster Video card and Dewitt Worrell's business card holder.

They also found three fragments of human bones.

15

STALLED

The phone rang just as Haueter and his wife, Trisha Dedik, were headed out on a Friday night. It was Ron Schleede, one of Haueter's bosses. Within minutes, Haueter and Schleede had a three-way call going with Greg Phillips, the systems group chairman. For two hours, they talked about the latest developments in the USAir case and where they should go next. So much for Haueter and Dedik's Friday night plans.

Dedik was livid. The crash had consumed her husband's life. It interrupted dinners and dates and weekend plans. It kept him at the office late and sent him out of town. He was obsessed by it. They would be having a perfectly nice conversation, and then she would see his mind drift away as he contemplated some damn theory about the damn accident. And then, right in the middle of *their* conversation, he'd say something about the crash, as if he had not been listening to a word she said.

She would reply, "I don't care." She was afraid the investigation was going to destroy little bits and pieces of the man she loved until there was nothing left.

People from the NTSB and Boeing called him at home, day or night. Everything was an emergency, even when it really wasn't. It reminded Dedik of the story of the little boy who cried wolf. These guys—and they were nearly all guys—were always crying wolf. Worse, they treated her like she was Haueter's secretary. It infuriated her. "I don't care about the office, I don't care about 427, I don't care about anything," she told him one night. "The victims are dead. There is nothing you can do about it. You know what? It's not going to make any difference whether you solve this today or tomorrow. There is nothing that is so important that you have to deal with it right now."

She had an important job, too, keeping the world safe from a nuclear disaster. She worked hard and traveled the globe to negotiate with other countries and make sure they were complying with international agreements. But she didn't obsess about her job, and she knew how to draw a line between work and the rest of her life. When she left the office, she left her job behind. That was one of the things that she had liked about Haueter when they met—he had a life outside his job. He was different from the other Washington men. But Flight 427 had changed him.

Some nights when he arrived home, she would tell him not to say a word about the investigation. Other nights she would give him ten minutes to talk and then make him promise not to bring it up again.

She asked him once, "Is it so awful to have an investigation unsolved? Does that mean you're a failure?"

"Yes," he said.

Some people in aviation, and a few at the board, thought that too much emphasis was being put on coming up with the probable cause. To them, the NTSB frequently got tied in knots trying to find the precise cause of a crash and did not focus enough on preventing future accidents. The reformers, led by retired NTSB aviation safety chief C. O. Miller, wanted the board to issue "findings" after each crash, with the emphasis on preventing accidents and reducing hazards.

But to Haueter and most others at the NTSB, it was important to name the probable cause. They thought that the public needed answers about why a plane crashed. It gave everyone a sense of closure, and it gave people confidence that it was safe to fly. If the board didn't name a probable cause in USAir 427, people would wonder for years whether the 737 was safe.

Haueter was a long way from solving the mystery, however. It was nine months after the crash, and the investigation had stalled. They'd done hundreds of tests on the rudder system and come up empty. Many other theories had fizzled. Several colleagues at the safety board thought Haueter should admit he was stumped and give up.

With no new leads, Haueter went back to old theories that had been ruled out. It was a sign of how desperate he had become. Leads that had been dismissed nine months earlier were alive again. He was willing to consider anything, even if it did not involve the rudder.

Brenner and operations group chairman Chuck Leonard conducted a second interview with Fred Piccirilli, a witness on the soccer field who thought he saw smoke coming from the plane before it hit the ground. The wreckage showed no sign of an in-flight fire, but Piccirilli had credibility because he was a USAir maintenance employee. He had come up to Hall at the hearing in Pittsburgh and told him about the smoke, so Hall asked the investigators to talk to him one more time. In the second interview, he again said the smoke was "orangish-reddish-brownish" and was coming from a spot in front of the right wing. He saw no fire but said the smoke remained in the air after impact and dissipated slowly. His account hadn't changed much from the first time. The problem was, there was not a single piece of evidence that backed up what he said.

The bird theory had also been born again. Never mind that no one had found a single feather in the wreckage or that Roxie Laybourne had conclusively ruled out the suspicious clump. The Boeing investigators had persisted about the theory, saying it was possible a bird had broken through the nose of the plane and hit a rudder cable. They wanted to do one final check of the wreckage with a black light to look for bird remains. This would be their last chance because the NTSB was about to release control of the wreckage to AAU, USAir's insurance company, which planned to put the pieces in crates and move them to a warehouse.

Supplee, the USAir mechanic, had been summoned back to the Pittsburgh hangar to help conduct the final bird check. He thought the whole exercise was a waste of time. They had already ruled out birds, yet Boeing persisted with any theory, no matter how absurd, that would clear the airplane.

When he walked into the hangar, Supplee noticed that the musty smell of chlorine and jet fuel had faded since his last visit six months earlier. The windows had again been covered so his team could work in darkness. They donned the ridiculous-looking yellow-orange glasses that supposedly made it easier to see the glow from bird remains. They had two black lights, so they split into teams and started on different sides of the wreckage. Supplee and a forensic expert from the Armed Forces Institute of Pathology went to the place where the leading edge of the wings had been partially reassembled and waved the light over it. Nothing glowed. They moved to other piles of wreckage. Supplee picked up each piece, and the guy from AFIP waved the light over it. Still no glow. They checked hundreds of pieces and did not see anything glow.

Suddenly they heard a shout from across the hangar, where a Boeing team member had a second black light.

"Guys! Look at this!"

Everyone hurried over. They could see a piece of wreckage glowing like a Jimi Hendrix poster in a dorm room. If this was a bird, it must have been a big one. How could they have missed it before? Then they removed their glasses and shone a flashlight on the wreckage. It wasn't a bird. It was the fluorescent paint they had used to draw grid lines at the crash site. They must have painted the wreckage by mistake.

They laughed and went back to the tedious job of inspecting each piece. A little while later, the AFIP expert noticed a slight glow on one piece. "Might have something here," he said. It was shaped like an X, about one and a half inches long and half an inch wide. Supplee could see it clearly through his glasses. They took a Q-tip and a solution and swabbed the area gently to get a sample. When they added a solution to test for organic materials, it glowed slightly.

But lab tests were negative. Once again, birds had been ruled out.

Some days for Haueter went like this:

9 A.M. Boeing calls and whines about the investigation.

10 A.M. ALPA calls and whines.

11 A.M. It is USAir's turn, followed by a second Boeing whining session after lunch. Haueter wanted to shout: "Give me a break!"

He believed in the party system, but on many days the parties behaved like children. In the absence of solid answers, they had retreated to positions that protected their own interests. Boeing and Parker Hannifin, the manufacturer of the PCU, saw no evidence that the rudder system had malfunctioned. They still wanted Haueter to spend more time scrutinizing the pilots. But ALPA and USAir saw no evidence that the pilots had made a mistake. They wanted Haueter to conduct more tests on the rudder system.

Haueter was frustrated by the lack of progress. Many clues pointed to the rudder system, but he and Phillips could not find any proof that it had failed. The power control unit not only passed every test, it seemed to have the strength of a superhero. So, in May 1995, Haueter decided they should step back and take a broader look at the plane, instead of just focusing on the rudder. Maybe they would discover some wild idea that would lead them to the solution.

He, Phillips, and Vikki Anderson, the lead investigator for the FAA, flew to Greensboro, North Carolina, and rented a car to drive to Winston-Salem, where USAir had a big maintenance hangar. The tan-colored hangar was big enough to hold three 737s for inspections that USAir called Q checks, the major overhauls where planes were stripped to their frames so mechanics could replace anything that did not meet FAA standards. A 737 got a Q check every 11,000 flight hours, or roughly once every three to four years. It was a great place to look for inspiration because everything in the planes was exposed. The 737s looked like they were naked.

A steady rain fell outside the hangar as they wandered around one of the stripped planes, looking at the maze of cables and wires that were normally hidden by the floorboards and the aluminum skin. They scrutinized the wing, trying to see anything that might have caused the smoke the witness had reported, and crawled into one of the baggage compartments. Phillips was struck by how much older planes smelled like coffee. They carried so many thousands of gallons of coffee over the years that it seemed to permeate everything. It was a familiar, comfortable smell.

They sat in the cramped baggage compartment and talked about the possibility that an obese passenger fell through the floor and landed on the rudder cable—an idea that had been dubbed "the fat guy theory." Would the fat guy have pulled on the cable enough to move the rudder? Probably not, but it might be worth testing, Phillips said. They looked at the plane's tail and discussed what might have happened if a bird had struck the vertical fin. Could that have turned the rudder?

USAir mechanics had tagged along to answer questions. "Do you have any ideas?" Anderson asked them. "Is there anything, no matter how far out, that you have noticed about the plane?" They had a few suggestions, but they were all theories that the NTSB had considered before. Anderson went up to the cockpit and sat in one of the seats. She stepped on the rudder pedals, turned the wheel, and moved the control column up and back. She was struck that it was such a simple airplane, about as sophisticated as an old VW Beetle.

She and Phillips went to the wheel wells, where the hydraulic lines converged and where the landing gear was stored in flight. A USAir mechanic who accompanied them was concerned about the vulnerability of the hydraulic lines there. If a tire blew out, it could do serious damage, he said. Boeing had once had screens that protected these fragile components, but the company had stopped using them.

"I've always wondered about things being in such a small place," the mechanic said.

"I have, too," Phillips said. But he had found no evidence of such an explosion in the wreckage.

They spent lots of time around the tail, looking at the power control unit and how it fit inside the vertical fin. Could a bird have broken through the tail and hit the cables or rods that led to the PCU? Haueter doubted it. The angle of the vertical fin would make it hard for a bird to break through. It would be deflected before it could pierce the skin.

Rain was pouring down as they had lunch at a pub near the hangar and then took a commuter flight back to Washington. It had been a good chance to see the innards of the plane, but they felt no closer to knowing what had happened.

Boeing had a theory about the pilots' feet. The company wanted to test the damaged pilot seat tracks to determine if they indicated where Emmett and Germano were sitting. The pilots might have been so far from the rudder pedals that they had to stretch to use them. Maybe that caused one of the pilots to push too long and hard on one of the pedals.

The seat tracks, metal strips that allowed the pilots to slide their seats forward and backward, were brought to the NTSB's metals lab in Washington. Metallurgists inspected the tracks, but they were so mangled that the experts could not find proof of whether the seats were too far back. Besides, the position of the seat was not important, since each pilot had a crank that adjusted the rudder pedals forward and backward, so even when a pilot had the seat back as far as it would go, the pedals could be in the proper place. At six-three, Emmett was so tall that it was logical that he would have his seat back.

Cox felt that Boeing was going to ridiculous extremes to blame the pilots. "Boeing is desperately trying to do anything they can to clear their airplane," he said. He was resolute that the pilots were not at fault. "I still think it's a systems problem with the airplane."

A few days later Cox and the other investigators flew to the NASA-Ames Research Center near San Francisco to take a ride in one of the world's most advanced flight simulators. The idea—also suggested by Boeing—was to see if Germano and Emmett had been so startled by the wake turbulence that they could not recover the plane.

The NASA vertical motion simulator, or VMS, was part of SimLab, the world's most advanced laboratory for studying pilots. The lab gave out patches with the simulator's logo, which had four stars and arrows pointing in every direction. It looked like a recipe for motion sickness.

The VMS resembled a wandering elevator inside a cavernous ten-story building. Like Boeing's M-Cab simulator, it could be programmed to perform like many different aircraft—fighter planes, big transport jets, even the space shuttle. It felt more realistic than other simulators because the cab could go up and down sixty feet, compared with just a few feet on M-Cab. That gave pilots a better illusion of the ups and downs in a plane. For the Flight 427 test, airsickness bags had been taped to the walls, just in case.

Malcolm Cohen, a NASA expert on the inner ear, had been invited to determine if the pilots had lost their bearings when the plane started to flip. He rode twelve times—several with his eyes closed—and was surprised at how smooth the ride was. He said it was so smooth that the pilots should not have been disoriented. Cox had watched the simulator go back and forth a few times before his turn and said it would be "an E-ticket ride," like the most thrilling ones at Disney World. He then walked across the ramp, buckled himself into the left pilot's seat, and put on the headphones. The safety board had brought along the cockpit tape so riders could listen to the crew.

The cab moved down to one side of the big tower, as if the NASA people were getting ready to fire it from a slingshot. The ride began. It was much smoother than M-Cab, without the jerks of the mechanical stops. Again and again, Cox rode the simulator, concentrating on a different element each time. He watched the instrument panel on some rides and looked out the windows on others. The simulator's windows showed a computerized display of the ground and sky that changed as the plane moved.

He had seen a similar display in Boeing's M-Cab, but he noticed something new this time. He was surprised at how quickly the plane went nose down and the windscreen filled with the ground. It seemed like the pilots were helpless to do anything. To Cox, it was more evidence that the crew was fighting to survive, but the plane was not responding.

Haueter hardly ever got scared when he flew, but he did get the jitters during a flight in a USAir plane in the summer of 1995. The plane was just like Ship 513, a 737-300. It was approaching Washington National Airport when it hit turbulence. Startled, Haueter sat up straight and quickly looked around to find the horizon through the windows, to see if the wing had rolled down too far. What was going on? Had the plane had a rudder hardover? Unfortunately, the people beside him had pulled down the window shade. He looked out the windows on the other side, but he couldn't tell how far the plane had rolled. A quick thought flashed through his mind: *Wouldn't it be ironic if the chief investigator for 737 crashes was killed in one?*

The plane leveled off, and Haueter breathed a sigh of relief—but he vowed to get a window seat in the future.

A big reason for his jitters was that he felt the 737 needed immediate safety fixes. They had not come up with the probable cause, but he and Phillips felt they had found enough problems with the rudder system to ask the FAA to mandate some improvements. That was typical in a high-profile investigation. The NTSB often made safety recommendations long before it determined the probable cause. Phillips had written a fifty-page memo calling for eighteen safety improvements, most involving the rudder. The memo said the FAA should immediately require Boeing to devise a procedure for pilots to handle a rudder hardover. Phillips was concerned that pilots would be caught by surprise if they had an incident and wouldn't know how to respond. The memo also called for long-term design changes to the PCU to prevent a hardover.

In response to Boeing's inability to track rudder problems with its databases, Phillips also called for a joint program between government and industry to keep a database on maintenance and operational problems. He had been surprised that a company as sophisticated as Boeing could not easily list 737 rudder incidents.

Haueter thought the memo made a good case for the changes. He also wanted to get a jump on the FAA, which was writing its own safety study on the 737 and relying heavily on what Phillips had learned. Haueter did not want the FAA to get credit for the NTSB's work. "These guys are going to beat us to the punch with *our* data," Haueter told Bud Laynor, the NTSB's deputy chief of aviation safety.

Laynor was a navy-trained pilot and engineer who had designed flight control systems on airplanes and NASA spaceships. At sixty-one years of age, he was in great physical shape, with short brown hair cut like he was still at the Patuxent River Naval Air Station and a thin, leathery face with creases in the cheeks. Laynor was the most respected technical expert at the board, so his approval for Phillips's memo was crucial. Without it, the memo would go nowhere.

When Phillips wrote it in the spring of 1995, he figured it would go through the usual editing by NTSB managers and would emerge essentially intact. But after a few weeks he was surprised that nothing had happened. The memo seemed to be stuck on Laynor's desk. When he saw Laynor in the hallway one day, Phillips mentioned the memo. "What about those recs?" Phillips asked.

"I'll get to them," Laynor said.

But when he did get to them, Laynor decided the memo was premature. The 737 had more than 60 million flight hours, and not even one crash had been linked to the alleged rudder problem. The NTSB had no evidence that anything in the PCU had malfunctioned. Without proof, he felt it was premature to say there was a problem with it. Besides, the valve had a built-in backup. If one slide jammed, there was a second slide to oppose the jam so the rudder could still move or return to neutral.

Laynor didn't argue much against Phillips's memo, he just sat on it. The memo stayed on his desk, gathering dust. Phillips, normally one of the NTSB's most cautious, gentle investigators, complained loudly to everyone that the fixes were crucial. These were not a bunch of crackpot recommendations, he said. There were persuasive arguments for each of them.

Haueter agreed. He felt so strongly about the safety fixes that he went over Laynor's head to Hall. But Hall was unwilling to get involved. If Laynor wasn't ready for the recommendations, no one was. The 737 safety fixes would have to wait.

Laynor was still interested in theories about the wake turbulence from the Delta plane. He said it was too much of a coincidence that Flight 427 had flown through the precise spot where the Delta plane had been seventy seconds earlier. He figured the wake had to play some role in the pilots' loss of control. He kept pushing for a flight test with a 737 and a 727 to show whether the wake was strong enough to flip a plane.

To Haueter, it was ridiculous to think that a wake could flip a 737. He repeatedly said that planes would be falling out of the sky every day if that were true. But he knew that he had to test Laynor's hypothesis. If he didn't, the wake turbulence theory would haunt them for years.

The FAA had its own 727 that could be used for the tests, but Haueter found it surprisingly hard to get a 737. They were the best-selling jets in the history of aviation, so he figured that some owner somewhere would be willing to lease one for a month. But he kept getting turned down. The 737s owned by the major airlines were all booked, and they couldn't spare one for a month, even though the NTSB was going to pay. Leasing companies had some available, but they didn't want to help. They were afraid the plane might be damaged by the test and did not want the publicity linking their plane to one that killed 132 people. Haueter offered to paint it white so that no one would know the owner, but he still couldn't get any takers.

Finally, he went to Hall and said he needed help. The test was worthwhile, Haueter said, but no one wanted to participate. Could Hall exercise some leverage and find a plane?

Hall got on the phone with FAA administrator David Hinson, and they jointly called USAir chairman Seth Schofield. Suddenly USAir changed its tune. The airline would loan a 737.

Everyone agreed that the $1 million cost of the tests would be split by the NTSB, the FAA, Boeing, and USAir. ALPA decided it could not afford to contribute, which Haueter thought was a bit cheap. Here's a bunch of guys making $160,000 a year, but they couldn't afford to support the tests?

28 seconds to impact, altitude 6,000 feet.
"Oh, yeah, I see zuh Jetstream."
"Sheeez."

15 seconds to impact, altitude 5,800 feet.
"What the hell is this?"

The View from the Cockpit

Using information from the flight data recorder and the radar track, photographer Bill Serne re-created the final 28 seconds of Flight 427. He shot the pictures from a helicopter matching the altitude and pitch to show what the pilots would have seen out their window as the jet spiraled down. A note about altitude: The plane crashed into a hill that was about 1,300 feet high, so impact occurred at 1,300 feet, not zero.

(Photographs copyright 1999 *St. Petersburg Times*)

12 seconds to impact, altitude 5,000 feet.
"Oh God . . . Oh God."

2 seconds to impact, altitude 1,800 feet.
"God!"

`00:00:01`

1 second to impact, altitude 1,500 feet.
"Noooo . . ."

Joan Van Bortel. She left for the airport so late that coworkers thought she would miss her flight.

(Photo courtesy of Brett Van Bortel)

Brett and Joan Van Bortel at a company party on September 2, 1994, six days before the crash. Joan had lots of ambition. She wanted to be the highest-ranking woman at Akzo Nobel.

(Photo courtesy of Brett Van Bortel)

The grisly scene on the hill. Investigators had to sift through dirt and vegetation to find the tiny pieces of wreckage.

(Photo courtesy of the FAA)

Plastic suits. Investigators had to wear protective clothing and be decontaminated with a bleach solution.

(Photo courtesy of the FAA)

The scene in the Pittsburgh hangar. The wreckage was laid out in the shape of an airplane.
(Photo courtesy of the FAA)

An impossible jigsaw puzzle. Investigators found a few large pieces of the wing and fuselage, but most were no bigger than a car door.
(Photo courtesy of the FAA)

NTSB chairman Jim Hall amid the wreckage.

(Photo courtesy of the FAA)

A piece of Ship 513's fuselage. The plane virtually disintegrated when it struck the hill.

(Photo courtesy of the FAA)

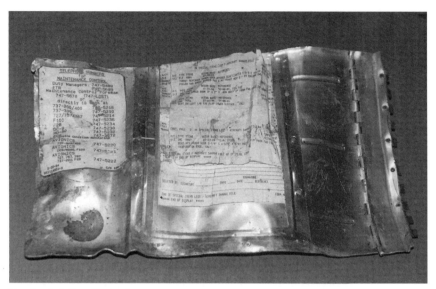

The plane's battered logbook, recovered from the wreckage. The log showed that Ship 513 had no significant mechanical problems before the flight.

(Photo courtesy of the FAA)

The mangled flight data recorder. It took only thirteen measurements and could not answer the fundamental question of the investigation: Did man or machine cause the crash?

(Photo by Bill Serne; copyright 1999 *St. Petersburg Times*)

Tom Haueter. He brought a refreshing dose of Midwestern charm to the investigation. "Holy mackerel!" he frequently exclaimed.

(Photo by Bill Serne; copyright 1999 *St. Petersburg Times*)

Jean McGrew. Boeing's chief engineer for the 737 believed the pilots were startled and caused the crash.

(Photo by Bill Serne; copyright 1999 *St. Petersburg Times*)

Greg Phillips holding the unique rudder valve from the USAir plane. He was the most cautious voice at the NTSB. While other investigators were convinced there was a malfunction, Phillips was still unsure.

(Photo by Bill Serne; copyright 1999 *St. Petersburg Times*)

A ghost ride in Boeing's M-Cab flight simulator. M-Cab rides allowed investigators to re-create the final seconds of Flight 427 and feel the same bumps and twists that the pilots did.

(Photo by Bill Serne; copyright 1999 *St. Petersburg Times*)

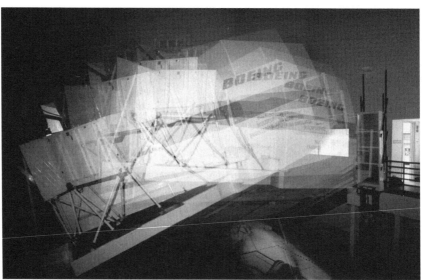

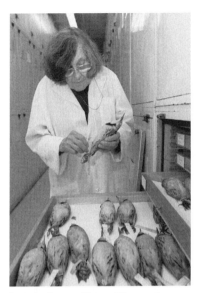

Roxie Laybourne. The world's expert on feather identification was called in to determine if a bird caused the crash.

(Photo by Bill Serne; copyright 1999 *St. Petersburg Times*)

John Cox. The USAir pilot believed there was "a gremlin" in the plane that made the rudder malfunction.

(Photo by Bill Serne; copyright 1999 *St. Petersburg Times*)

Brett Van Bortel at the crash site on the first anniversary. He looked up at the sky as a USAir 737 passed overhead.

(Photo by Bill Serne; copyright 1999 *St. Petersburg Times*)

Mementos in the woods. This tree, photographed on the one-year anniversary, still bears scars from the crash.

(Photo by Bill Serne; copyright 1999 *St. Petersburg Times*)

The investigators talk with the press before the flight tests. From left, Bud Laynor, Mike Benson, Tom Haueter, and Tom Jacky.

(Photo courtesy of the FAA)

The wake turbulence test. Pilots flew the USAir 737 through the wakes of another plane to see if turbulence played a role in the crash.

(Photo courtesy of the FAA)

Cox (visible in the cockpit) flies the USAir 737 during the tests. A T-33 chase plane flies beside the 737 to take photos.

(Photo courtesy of the FAA)

Haueter in his Stearman biplane. To get a break from the pressures of the investigation, he went flying.

(Photo by Bill Serne; copyright 1999 *St. Petersburg Times*)

Testing the fat guy theory in a Boeing hangar. The investigators wondered if an overweight passenger had stepped through the floor onto a rudder cable.

(Photo by Bill Serne; copyright 1999 *St. Petersburg Times*)

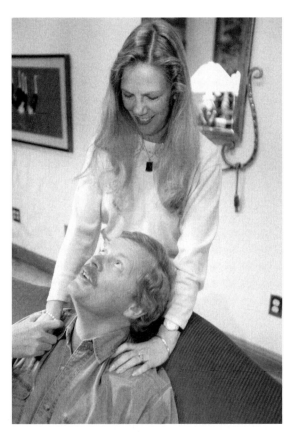

Tom Haueter and his wife, Trisha Dedik. She grew frustrated as the investigation consumed her husband and interrupted their life together.

(Photo by Bill Serne; copyright 1999 *St. Petersburg Times*)

A memorial to the crash. The names of Flight 427's 132 passengers are etched in granite at the Sewickley Cemetery outside Pittsburgh.

(Photo by Bill Serne; copyright 1999 *St. Petersburg Times*)

JOSEPH R. KOON
JOHN J. KUPCHUN
CAROLYN KWASNOSKI
DANIEL KWASNOSKI
JOAN E. LAHART-VAN BORTEL
IN DAVID LAMANCA
LOTTE L. LANGAN
LANGAN

Brian Bishop. The Eastwind Airlines pilot experienced a rudder incident that was eerily similar to that of Flight 427.

(Photo by Bill Serne; copyright 1999 *St. Petersburg Times*)

Malcolm Brenner. The NTSB human-performance expert studied the pilots' grunts and cursing for clues about what happened.

(Photo by Bill Serne; copyright 1999 *St. Petersburg Times*)

16

THUMPS

Jim Cash had the creepiest job in aviation. As the NTSB's expert on cockpit tapes, he listened to pilots die every day. He estimated that he had heard about three hundred die horrible deaths, plus several hundred more who had been fortunate enough to survive. It was fitting that Cash was shy and soft-spoken. He heard pilots hollering and screaming every day, but when he talked, his voice was so quiet that you had to strain to hear him. He was forty-three, with rosy cheeks and a boyish face. There were no aviators in the Cash family when he grew up in upstate New York—his dad had a blue-collar job at a power plant, his mom was a housewife—but Cash built model airplanes and got interested in flying. He was awarded an ROTC scholarship to Syracuse University and studied electrical engineering. It wasn't that he really wanted to be an engineer, it was a matter of survival. He hated writing and liberal arts and figured his best hope for graduating was to be an engineer. He joined the air force and flew F-4 fighter planes, but you would never know it by talking with him. He had none of the ego or bluster of a fighter pilot. He was in the air force for eight years but chose to leave when he was due for a desk job.

He couldn't find anything he liked at an aerospace company and stumbled across an opening in the NTSB sound lab. He had no experience listening to cockpit tapes (air force planes usually do not have them because the military does not want the tapes to fall into enemy hands), but his experience as a pilot and his skills as an electrical engineer made him ideal for the job. Like Haueter and Phillips and seemingly everyone else at the safety board, Cash loved to build and fix things. He built a pool in his backyard and spent four years restoring a '72 Porsche 914.

Cash was renowned for finding clues in the tapes. He knew the distinctive clicks of a flap lever and the grind of the landing gear. By studying the whines of the engines, he could tell if they were operating at full power. His office at the NTSB was a darkened room that was stacked high with audio equipment—amplifiers, tape players, audio mixers, and graphic equalizers. But the place where he worked his magic was a powerful Unix computer that ran a program called WAVES. It allowed him to translate sounds into squiggly lines that showed frequency, volume, and energy. When he wanted to identify a strange click or rattle, Cash compared that fingerprint with one from a known sound. He spent lots of time recording sounds on airplanes on the ground so he could match them with the mysterious noises. He was like a police fingerprint expert, always looking for a match.

It had been ten months since the crash, and the NTSB still had not identified the thumps heard on the 427 cockpit tape. Cash had taken the tape to the FBI's sound laboratory in Quantico, Virginia, to see if they came from gunshots or an explosion, but the FBI found that the thumps did not have the unique signatures of sounds from a gun or a bomb. Since then, Cash had spent hours studying the thumps on his computer. He discovered that there actually were three distinct sounds, the first coming just as Emmett completed the sentence "Oh, yeah, I see zuh Jetstream." But Cash had been unable to match the signature with any of the sounds from the stomping-slamming doors test conducted at National Airport a few days after the crash.

If the thumps had been heard at any other time on the cockpit tape, Cash would not have worried about them. Tapes often had sounds that couldn't be identified. Human beings and airplanes made so many noises that it would be difficult to identify them all. But Cash thought it was important to find the source of the thumps because they occurred at the most crucial time, right before the plane rolled left and twisted toward the ground.

It was fortunate that he had four channels of sounds on the tape. Most 737 tapes had only three—one for each pilot and one from the area microphone in the ceiling. But someone on Ship 513 had mistakenly left the jump seat mike on, which meant there was a fourth source of sounds.

As he studied the squiggly lines on his computer one day, Cash noticed that the thumps were picked up by the microphones in the ceiling and the jump seat but not by the headset mikes that Emmett and Germano were wearing. That meant the sound was probably traveling through the frame of the airplane rather than through the air, which suggested it came from outside the cockpit. He also noticed an important difference in the sounds: the ceiling mike picked up the first thump $9/100$ of a second before the jump seat mike did. He thought that was an important clue.

Late one summer night, Cash and a few helpers arrived at the USAir terminal at National Airport. They boarded a 737 parked at a gate and switched on the plane's power. As the voice recorder was running, an FAA employee crawled through the baggage compartment and pounded on different spots with a rubber mallet.

They used two walkie-talkies. The FAA employee announced his location right before he pounded, so they could log where the sounds came from. When Cash got back to his office and examined the sounds in WAVES, he checked the timing of each to see which ones had the same $\frac{9}{100}$ of a second interval. He found a match when the FAA employee pounded on the fuselage about twelve to sixteen feet behind the cockpit mike. That meant the sound came from Row 1 or Row 2 in first class.

He was halfway there. He knew the location of the thumps. Now he just had to figure out what caused them.

Every time a 737 burped, Vikki Anderson's phone rang. It didn't have to be anything serious. The usual hiccups of a plane—minor autopilot problems, rudder anomalies, or bumps from wake turbulence—were now being reported by 737 pilots as potentially serious incidents that might be related to the USAir crash. Many pilots had become jittery about the 737.

As the FAA's lead investigator on the crash, Anderson provided technical help to the NTSB, to explain how the FAA certified the 737 and how it inspected USAir. She also had to help the FAA decide if there was any need to take action because of the crash. She had been a Braniff flight attendant for twenty-three years and had joined the FAA because she had grown weary of the vicious cycles in the airline business, when Braniff would go bankrupt and then reemerge to fly again. She talked about the airline like it was a movie with sequels, *Braniff I, Braniff II,* et cetera.

She started with the FAA as a cabin safety specialist, analyzing flight attendant training and evacuation plans, and then moved to the agency's accident investigation office nine months before the Hopewell crash. This was her first major accident. As she was flying to Pittsburgh the morning after the crash, she wondered if she could handle the gruesome job. She wasn't sure she could deal with body parts lying everywhere. She had done well in the months since then because she was forthright about her feelings. While many men in the investigation bottled up their emotions, Anderson was honest about what she was going through. She was a fun person to work for, easygoing, thorough, and well organized.

Only a handful of women were part of the tight fraternity of crash investigators, but Anderson had worked hard to learn the ropes. She read every crash book she could find, and she wasn't bashful about asking questions. She would freely admit when she was baffled by an engineer's mumbo jumbo. She kept the FAA team organized and made sure meetings didn't drag on too long. A tiny woman with big brown eyes and a warm sense of humor, she could liven up the most dreary meeting. When they discussed the gurgling sound that Andrew McKenna had heard in Seat 1A, Anderson suggested it was someone sitting in 3A gargling. When she visited the Winston-Salem hangar, she said she was open to all theories about the crash, "even the one about Russian death rays."

The FAA had declared the 737 safe. An independent team of FAA engineers and inspectors had conducted a Critical Design Review of the plane for six months and found no major problem that could be linked to the accidents in Hopewell and Colorado Springs. The team listed twenty-seven recommendations to improve the plane, everything from new screens in the wheel wells to a consistent definition of the word "jam," but it found no serious problems with the plane.

Yet, despite the FAA's assurances about the plane's safety, pilots remained nervous. In cockpits and airport hallways, 737 pilots eagerly swapped gossip about recent incidents and discussed what they would do if their planes suddenly began to roll. At least once a month, Anderson would get a call from the FAA security guard saying there was a pilot at the front door who wanted to talk to someone about the crash. Sometimes the pilots had theories, other times they just wanted to find out the latest from the investigation. One young pilot—Anderson doesn't remember his name or his airline—was surprisingly nervous about the 737. He needed reassurance that the plane he flew was not going to fall out of the sky. Talking to him was eye-opening. It made her realize that some pilots were truly scared.

Anytime a pilot reported a problem that sounded the least bit serious, the FAA and the NTSB quickly sent a team to check it out. A July 1995 incident in Richmond, Virginia, was typical. Anderson's boss, Bud Donner, interrupted her during a meeting and said, "Vikki, I'm going to need you in a little bit. We've got another little adventure."

A USAir plane carrying forty-eight passengers was making a left turn toward the runway when it suddenly rolled to the right. The captain shut off the autopilot, but the roll continued. He quickly turned the wheel and pushed the left rudder pedal to regain control of the plane. The rolling stopped, the plane leveled off, and they landed safely.

Anderson and Donner discussed the event and decided it sounded serious enough to send people to Richmond. FAA investigator Jeff Rich quickly rented a car and raced down I-95 to the airport. The NTSB also dispatched someone.

The plane was still parked at the gate when Rich arrived. For three days, he and other investigators crowded into the cockpit and ran tests on the plane's navigational computer, programming the computer so it thought the plane was flying and then checking to see how it might have sent a faulty signal. In the meantime, the plane's flight recorder was sent back to Washington to be analyzed in the safety board's lab. Ultimately, the roll was blamed on a faulty autopilot and on the pilots themselves, who did not realize the autopilot was steering the plane.

Two weeks later, Anderson's home phone rang while she was in the shower bathing her cocker spaniel, Pepe Lopez (a dog her kids named after a brand of tequila). An Aviateca 737 that originated in Miami had crashed into the side of a volcano in El Salvador, killing all sixty-five people on board. The plane was making a final approach to San Salvador during bad weather. Early reports suggested that there weren't many similarities with Flight 427, but no one could be sure. Clues to the Hopewell crash might be lying on the side of the volcano. Anderson packed her bags and flew to El Salvador.

Her first view of the crash came from the TV in her hotel room, which showed grisly pictures that would never be shown in the United States. The bodies were relatively intact, which told her this was a much different accident from Pittsburgh. The plane had probably struck the volcano with a glancing blow. The next day she got a Jeep ride as far up the 7,000-foot volcano as possible, but she had to climb the rest of the way in her blue FAA jumpsuit, wearing a 30-pound backpack loaded with tools, water, cameras, and biohazard gear. She had thought she was in great shape, a three-times-a-week runner, but the volcano was so steep that she had to crawl the last 50 feet on her hands and knees. When she reached the 6,400-foot level, where the plane had crashed, she lay on the ground gasping for breath, then looked up and saw an eighty-year-old Salvadoran woman grinning at her.

Anderson figured the woman had probably passed her going up the mountain. Maybe Anderson wasn't in such good shape after all.

Anderson and investigators from Boeing, the NTSB, and the Salvadoran government climbed through the jungle and inspected the wreckage. The voice recorder was found and was sent back to the NTSB lab, but the plane's data recorder was missing. Anderson figured a villager who lived near the volcano had probably stolen it, thinking it was some kind of safe box with money inside. That kind of looting was common at crashes in Third World countries. Investigators could often tell they were nearing a crash site because they started seeing people sitting outside their homes in airplane seats.

After analyzing the wreckage and the voice recorder and interviewing an air traffic controller, the investigators decided that the pilots were at fault. They had approached the airport from the wrong direction and had flown into the side of the volcano. The Richmond and El Salvador accidents ended up on a list with more than fifty other 737 incidents that were explained by mechanical malfunction, pilot mistakes, wake turbulence, or bird ingestion. They did not answer the riddle of Flight 427.

17

THE ANNIVERSARY

The Flight 427 Air Disaster Support League was making progress in its effort to improve treatment of families after a crash. On June 20, 1995, members of the league and three other crash groups met with Chairman Hall and Secretary of Transportation Federico Pena at the NTSB offices in Washington. The families from the other crashes had the same complaints as those from Flight 427: It took too long for the airline to figure out who had died, the coordinators for the airlines were inadequately trained, and the companies did a poor job of returning personal belongings.

Aides to Pena and Hall scribbled notes as the families listed their complaints. One of Pena's aides kept reminding him to leave for another appointment, but he stayed for an additional forty-five minutes. He and Hall came away convinced that the government needed to help. They directed staff members to draw up legislation creating a new government office to assist families after a crash.

That summer, the Flight 427 Air Disaster Support League was organizing two days of events to mark the first anniversary of the crash. The group planned to have four memorial services, including one at the crash site on the night of the anniversary, and a luncheon for the families the following day.

Brett had not been active in the league. He agreed with its goals, but he wanted to keep his distance. When he and Joan's brother Dan Lahart arrived in Pittsburgh for the anniversary, they stayed at the same hotel as other league members did, but they planned to skip the meetings and all but one of the memorial services. When they ate dinner at the hotel restaurant the night before the anniversary, they sat in a dark corner so they could have privacy.

Brett was hardly thrilled to be there, but he wanted to pay tribute to Joan and attend a memorial service at the place where she had died. He had endured the most painful days in the past year—his birthday, Christmas, her birthday—and hoped they wouldn't hurt as much in the future. He had not dated anyone since he lost Joan and was not sure he ever could. "If I meet someone in the future that I want to marry, she will know deep down inside that this would not be happening, we would not be having children, had I not lost the first love of my life," he said.

Brett and Dan visited the coroner's office in the morning, to check whether any of the unclaimed shoes or jewelry belonged to Joan. None did. Then they drove to Sewickley Cemetery to see the monument to the passengers. It was a beautiful day, just like the day Joan died. Workers were planting 132 tulips that would honor the victims. Brett and Dan got choked up. They chatted with a woman whose husband had been on the plane. She said she often visited the memorial in the early morning and found the cemetery very peaceful.

But Brett didn't find it the least bit peaceful, for it was directly in the flight path to the Pittsburgh airport. As he gazed at a memorial to the victims of one of the worst plane crashes in the nation's history, USAir planes roared overhead. He felt as though they were taunting him.

Gradually, George David was getting his property back. It had been a year since he heard the Boeing 737 roar through the air and crash on the edge of his 62-acre farm, but he still found remnants of the crash. Despite the huge cleanup effort, no one from USAir, the NTSB, or Beaver County had bothered to remove the hundreds of yards of red police tape around his property that was stamped DANGER—HAZARDOUS MATERIALS. On misty days he could still smell the aroma of jet fuel. His farm had once been a great place to hunt deer, but most of them had been scared away by the men in plastic suits.

In the year since the crash, David, a police officer and part-time hay farmer, had watched a parade of cars go down his dirt road, ignoring a dozen NO TRESPASSING signs. He didn't mind it if the visitors were families of the victims or USAir flight attendants who wanted to see where their coworkers had died. He had even hung a wreath in a tree for one of them. But he hated the souvenir hunters and the gawkers. Once, a woman walked up to him shouting, "Hey! Hey!" as he was quietly tracking a deer, which scared away the animal. Another guy had the nerve to come up to David's door and snap a picture of David and his fiancée watching TV. What could anyone possibly want with that picture? Another time, he got into an argument with a stranger who refused to leave. Finally, David slugged the guy and he ran away.

David and the two other families that owned the crash site felt they had been considerate for a whole year, allowing hundreds of people to visit and leave crosses and wreaths on their property. But they were ready to reclaim their peace and quiet. They decided no more visitors would be allowed after the anniversary. "I've got to get my life back," David said.

Brett met David during a visit to the crash site a few hours before the candlelight memorial service. "I'm sorry about your loss," David told him.

Brett had never spoken with anyone who had seen or heard the crash, and he was curious about what David remembered. He wanted to know what the plane had sounded like, whether David had seen it, how soon he'd gotten to the wreckage.

"It was roaring," David said. "It sounded like a Mack truck coming down a hill."

The small clearing that had been littered with airplane wreckage and body parts a year earlier was coming back to life. Grass was sprouting. Leaves had grown back on many of the trees, although some remained scarred by the crash and the fire, stripped of any sign of life for fifteen or twenty feet up the trunk. More than a dozen wreaths and mementos had been placed around the site. A crucifix was tacked on a blackened tree trunk. A cross was planted in the hill like a tombstone for passenger Leonard Grasso. Joanne Shortley had left a love message for her husband, Stephen, that said, "JS + SS" with a big heart, like something she would have written on a locker in junior high. Brett walked around the crash site searching the ground for the diamond from Joan's engagement ring. He knew it was silly to think that it might still be lying out in the open after a year, but stranger things had happened. This was the place where Joan's rose had been blown off the wreath, after all. Brett looked up several times at the USAir planes that flew over the crash site, about one per minute. They were probably at the same altitude as Joan's plane had been, filled with people who never considered that life might suddenly end.

He and Dan returned to the hotel to get dressed and then went back to Green Garden Plaza to get a ride to the memorial service. They were in one of the first buses that crunched up the gravel road and stopped near the site. Brett got out and walked down toward the place he called ground zero, where the plane's nose had hit the road. A few people crouched along the road arranging candles in the shape of hearts. Brett walked back up the hill as several hundred family members gathered around the Reverend Thaddeus Barnum, the leader of the service. Brett stood right behind a tree where a baseball cap had been posted in honor of one of the passengers. The hat was labeled SUPER GRANDPA.

A 737 happened to fly overhead at 7 P.M., just as the service began. Brett looked up at it. "Let there be silence among us," said the pastor as the plane's roar faded in the distance. At 7:03, the precise moment of the crash, church bells rang throughout Pittsburgh in honor of the victims. Everyone at the service said the Lord's Prayer and recited the Twenty-third Psalm. Then the pastor said, "I know many people could not come back to this site. I am very glad that you did, to come and know and feel and remember and touch and taste and to feel again, the memory of those you've lost."

That night, as he ate dinner at a Mexican restaurant near the hotel, Brett felt relief that he had gotten through the anniversary. He had experienced a lot of anxiety leading up the ceremony, and now he was glad to move on. But the next day the sadness returned. Life seemed a dull gray once again. At times he would forget about the crash, when he would be having fun with his brothers or his mom or his friends. But then he would suddenly think, *How can I be sitting here having a good time when Joan is dead?* A friend called them "grief bombs." When one exploded inside him, he lost track of what he was saying. He became quiet and often excused himself to get some fresh air.

Three days before the anniversary, USAir surprised Wall Street analysts by announcing that it would make a pre-tax profit for the year. That was a remarkable turnaround for a company that had seemed so close to filing for bankruptcy a year earlier. The company's stock, which had sunk to $3\frac{7}{8}$ after the crash, had rebounded as high as 14 in June 1995.

The rebound had less to do with what USAir did than what others had done. Continental Airlines, which had invaded USAir's East Coast stronghold a year earlier with its low-cost "Cal-Lite" service, pulled out in the spring, which allowed USAir to raise ticket prices enough to be profitable again. Most important, the public forgot about the airline's crashes and safety problems. They were old news by the summer of 1995. With no new crashes or incidents, people again were content to put their lives in the hands of USAir.

18

THE HOLE IN THE
FLIGHT ENVELOPE

As John Cox walked up the stairs to the USAir plane at Boeing Field in Seattle, he saw the word "EXPERIMENTAL" in big letters over the door. The sign beneath it read, THIS AIRCRAFT DOES NOT COMPLY WITH FEDERAL SAFETY REGULATIONS FOR STANDARD AIRCRAFT. The USAir plane had been specially equipped for the tests. It had a flight data recorder that took thirty measurements, more than twice as many as the box on Flight 427. First-class seats had been removed and replaced with computers and video equipment. Seven tiny video cameras were installed in the cockpit, in the windows overlooking the wings, on the wingtips, and on the tail. The plane had been loaded with several tons of sand to simulate the weight of passengers and luggage.

Haueter did not expect any breakthroughs from the flight tests. He was doing them just to appease the parties and put to rest Laynor's questions about the wake. Yes, the wake had jostled the plane and may have initiated whatever went wrong on Ship 513. But it surely didn't flip the plane out of the sky. He was confident the tests would show that.

There would be two sets of tests, one in Seattle and one in Atlantic City, New Jersey. The Seattle flights were called the simulator validation tests, which was an effort to keep M-Cab honest. Much of the investigation had been based on M-Cab's computer estimates about how 737s behave. Some people—most notably the head of the FAA, David Hinson—wanted to be sure that M-Cab accurately portrayed real 737s. The tests would also give

---

144

the investigators a chance to learn about the crossover point, the precarious moment in flight when the plane was at the mercy of the rudder. It was an aerodynamic quirk that Cox called "the hole in the flight envelope."

At higher speeds, a plane would be going fast enough that the ailerons on the wings could easily counteract a sudden movement by the rudder. The pilots simply turned the wheel. But if the rudder suddenly went hardover at speeds slower than the crossover point, the plane would roll out of control unless the pilots knew exactly what to do. They had to gain airspeed quickly and turn the wheel fully the opposite way of the rudder. But pilots of 737s were largely unaware of the phenomenon.

When the plane was certified in 1967, Boeing told the FAA that if a pilot lost control of the plane because of a failure in the rudder valve, the problem could be countered by using the ailerons. The company later discovered that there were speeds at which that wasn't true, but the discovery was not regarded as critical. Boeing did not mention the crossover point in its flight manuals or alert airlines or pilots about it. The company did not believe the crossover point was a big deal.

Members of the NTSB aircraft performance group had noticed the crossover point during tests in M-Cab, but they needed data from a real 737 to determine precisely where it was. That information could help to settle the debate between Boeing and ALPA about whether the pilots could have prevented the crash.

The plane, which had a radio call sign of "Boeing 053" instead of a USAir flight number, took off into sunny skies the morning of September 20, 1995. It headed north over the Straits of Juan de Fuca, the spectacular coastline along the Canadian border. Cox, Boeing test pilot Mike Carriker, and USAir pilot Jim Gibbs each had a chance to feel the crossover point for himself.

When Cox got his turn in the cockpit, he pulled the throttle back to slow the airspeed to 190 knots, with the flaps set at "1," just as they had been for Flight 427. He steadily pushed his foot on the left rudder pedal and simultaneously turned the wheel to the right to keep the plane from rolling, a maneuver known as a steady-heading sideslip.

He flew for miles at that fragile point, balancing the wheel and the rudder pedal. If he slowed to just under 190, he started to lose control. He had to push the stick forward to lower the nose and gain airspeed to recover.

Something is wrong here, Cox thought. The crossover point was way too high. He had expected it to be down around 170 knots, but it was actually at 187. The other two pilots tried the same maneuver and had slightly different interpretations of the speed, but they agreed on an important fact: Flight 427 was right at the crossover point when it flipped out of the sky.

Haueter, who had stayed in Washington, got an excited call about the discovery from Tom Jacky, the NTSB engineer who headed the performance group.

"They ran out of roll authority," said Jacky

"You're kidding," said Haueter.

"If you slow the airplane up, you don't have enough wheel to stop the roll."

"Holy shit!" Haueter said.

With the Seattle tests complete, Cox and the other pilots flew the plane across the country to Atlantic City for the wake turbulence tests. The FAA had agreed to loan its 727, a plane used primarily for research, to play the part of the Delta plane that was four miles ahead of Flight 427. The FAA jet had been equipped with special smoke generators on its wings.

Wakes were usually invisible, but the 727's smoke generators made them show up as spinning white tubes. They twisted across the sky like an abstract painting. The dual wakes would keep spinning for minutes, and depending on the winds, they would turn, curl, and bounce off each other. Standing near the runway when the FAA plane flew past, Cox could hear the *whooosh* from the spinning tube of air as it lingered long after the plane was gone.

The next day, Cox got to see the phenomenon from the air as his 737 flew in and out of the 727's wakes. They were stronger and closer together than Cox and the other pilots had expected. They marveled at the way the lines twisted and soared in the sky. It was as if they could finally see the pothole that they had been running over all these years.

"Oh, that was neat," Cox said to Boeing pilot Mike Carriker in the cockpit after they watched the wakes twirl back and forth. "I liked that."

For a week they tried 160 different conditions behind the 727—flying across the wakes, going up and down through them, and holding the wings, tail, or engines in them. For several tests, they took their hands and feet off the controls and rode the wakes like a roller coaster.

As Haueter had expected, there were no breakthroughs. The pilots could easily recover from a wake, no matter the angle at which they flew into it. When Phillips asked Cox whether he was startled by the wake, Cox said he was not. Reacting to the wake, he said, "is as natural as breathing."

But the engineers learned a lot about wakes. They learned how long they lasted and how they could roll a plane. Haueter and Cox said afterward that the tests were well worth the $1 million price because the data could be used by researchers for years.

The biggest breakthrough of the Atlantic City tests involved the mysterious thumps that had been heard on Flight 427's cockpit tape. Cash, the shy sound expert, had rigged the voice recorder in the 737 so it was identical to the one on Flight 427, and he had Carriker fly in and out of the wakes.

Carriker and the other pilots heard whooshing sounds when they flew in the wakes, but the sounds were not consistent. When Carriker landed, he told Cash the sounds didn't match the thumps from 427.

Still, Cash believed he might find a match. Noises often sounded different on the tape because the cockpit microphone picked up sounds through the airframe. Cash took the recording back to his hotel room, put headphones on, and queued up the tape.

Bingo! Exactly the way it sounded on Flight 427! The thump was caused by the Delta plane's wake. He could finally rule out the theories about explosions and birds. He called Haueter the next morning. "We've got a really good match," he said.

The sparring between Boeing and ALPA had remained behind the scenes, confined to phone calls and closed-door meetings. The angry exchange of letters about pilot error had not been released to the public. But as the two rivals prepared for the second hearing in Springfield, Virginia, in November 1995, it appeared that the conflict would go public.

It was highly unusual for the NTSB to have a second hearing, but Chairman Jim Hall wanted it to release findings of the flight tests and give an update on the investigation. Hall wanted to reassure people that the crash had not been forgotten. He often said that the NTSB was working on three accidents: "Pittsburgh, Pittsburgh, and Pittsburgh." The safety board chose suburban Springfield for the hearing because it was the closest place to Washington that had an available ballroom.

Boeing and ALPA came to Springfield with sharply different goals. The union wanted to keep the focus on the plane and emphasize the crossover point, but it was walking a thin line. Its leaders did not believe the 737 was unsafe—Cox and hundreds of other union members still flew it every day—but they wanted to emphasize the plane's problems and show that the 737 needed safety fixes.

The union leaders expected Boeing to come out with guns blazing, raising questions about whether Emmett and Germano had made mistakes in the cockpit. So ALPA prepared two strategies to respond through the news media. A normal, subdued approach would be used if the debate was civil. But if the hearing turned into nuclear war, the union would use tougher words. Cox was not on the initial list to testify, but two days before the hearing ALPA leaders decided that he should. They wanted to give the union's account of the flight test to counter the presentations from several Boeing witnesses. The union was wary of John Purvis, Boeing's chief accident investigator, who could cross-examine witnesses like a crafty trial lawyer. The ALPA leaders were afraid that Purvis might try to back Cox into a corner by asking hypothetical questions and challenging Cox for not being a test pilot.

Boeing's goal was just the opposite from ALPA's: The company wanted to keep the focus on the possibility of pilot error. McGrew was convinced that Emmett and Germano had been startled by the wake turbulence, had made a crucial mistake by pulling back on the stick, and had failed to turn the wheel right and keep it there. Boeing witnesses would remind everyone that there still was no evidence that the plane had malfunctioned, emphasize that the wake turbulence was "an initiating event," and raise the possibility that one or both pilots had stomped on the rudder pedal. The word "startled" would be used a lot.

The hearing in the Hilton ballroom was again set up like a giant trial, with

Boeing attorneys and engineers clustered around one table, the ALPA team at another. The other parties in the investigation — USAir, the machinists union, the FAA, the valve manufacturer Parker Hannifin, and the flight attendants union — also had tables. But this would primarily be a showdown between Boeing and ALPA.

The hearing had a sideshow as well, a press conference by Philadelphia trial lawyer Arthur Wolk. A news release said that Wolk would "unveil the real causes of the Boeing 737 crashes that continue to baffle the NTSB." The investigators regarded Wolk as a lawyer eager for a headline. Most of the press corps skipped his news conference and went to dinner. A few reporters showed up, along with Vikki Anderson, the FAA investigator. She sat in the back row, dutifully taking notes. (After the news conference, she said the head of the FAA had asked her to attend and write down Wolk's theories. The FAA was open to suggestions from anyone, she said.)

Wolk lashed out at the party system, saying, "Boeing has been involved in every single aspect of this. It makes no sense to have the company that has the most to lose involved in the investigation." Using a model of a USAir plane, Wolk showed how the plane had rolled out of the sky. He said the PCU was faulty because the valve could reverse, but he could offer only the same theories the safety board had been chasing for months. The reporters asked a few questions, munched on free potato chips, and left.

At the hearing the next day, as Cox walked up to testify, USAir lawyer Mark Dombroff whispered to a company official, "Cox is about to dump on Boeing." But Cox was cautious. He explained his concerns about the crossover point and his surprise that it occurred at 190 knots/flaps 1, a routine speed and setting for an approach to an airport. "I would have expected more padding underneath," he said. He delicately questioned the accuracy of Boeing's simulator but said he was pleased that the company was going to correct it. He said he would not be disoriented by wake turbulence if he encountered it during a flight.

The Boeing witnesses had a different spin. Carriker, the test pilot, stuck to the company line that the soda can valve should prevent rudder hardovers. Boeing did not require any pilot training for a hardover, he said. "We don't train for events that don't occur."

When the parties got their turn, ALPA leader Herb LeGrow asked a pointed question to remind everyone that Carriker had never flown for an airline. That tactic was used by both parties at the hearings. Boeing and ALPA tried to land punches on each other and toss softballs to their own people. Carriker struck a nerve with ALPA when he referred to "average" pilots. That brought laughter from the room because everyone realized that in the union's view there was no such thing as an average pilot. They were like children from Lake Wobegon. They were all exceptional.

LeGrow also sparred with McGrew, the Boeing engineer, about the crossover point and the company's willingness to include ALPA members on a new Roll Team that was investigating reports of sudden rolls by 737s. Mc-

Grew said, "We at Boeing have offered to the parties and to the NTSB in the past to please send representatives at any time to come and sit with us as we go through this investigation. We would be happy to accommodate you."

The union official shot back, "Mr. McGrew, I've been the coordinator of this accident since September 8th of last year and I have received no such communication from the Boeing Airplane Company."

"Excuse me, Mr. LeGrow, but I made that same statement sitting at the stand back in January." Later, McGrew scoffed at the importance of the crossover point. "Our basic position today is that the airplane has proved its airworthiness over the years and that this is probably not a significant item," he said. Boeing was planning improvements to two rudder system components, but neither one appeared to have played a role in the Hopewell crash. "There were no faults found in the mechanical systems of the airplane," he said.

Once again, McGrew's self-confident manner came on strong. He got snippy when anyone suggested that anything was wrong with the airplane. Haueter questioned whether older 737s should be required to meet current safety standards, but McGrew said it wasn't necessary. "If you buy a toy wagon for your child, and it wears well and is still usable when he's your age and has [his own] child, should you go out and re-fit it again? It's perfectly functional and works, nothing wrong with it."

Haueter grinned and said, "I guess using your analogy, we wouldn't put airbags in cars nowadays."

McGrew retorted, "I think now we're getting into arguing the relative safety statistics of the automobile versus the airplane. And I think you'll lose."

Relations between Boeing and the safety board were getting rocky. NTSB officials said Boeing had not told them about the Roll Team until two days earlier—even though the team had been investigating the incidents for more than a month. Boeing had distributed a packet of lists and charts showing that many "suspicious" rolls reported by 737 pilots were actually encounters with wake turbulence. Boeing's point was that there was no gremlin in the plane and that pilots just overreact when they hit a wake.

Included in the Boeing packet was the Roll Team's list of 737 incidents and the company's conclusion on each one. The first twelve were attributed to wake turbulence, five were blamed on minor airplane malfunctions, and three on pilots. If Boeing had left Flight 427 off the list, the NTSB would not have made such a fuss. But there it was, listed as "Roll Event No. 2." The "Boeing conclusion" for the crash was listed as "wake turbulence."

That was heresy. Only the NTSB was supposed to determine a probable cause. It looked as if Boeing was saying it had solved the case while the safety board was still stumped. Yes, the wakes might be a factor, but they were not the primary cause. McGrew and other Boeing officials apologized when they testified, saying that the Roll Team was a sincere attempt to help the airlines. They said it never crossed their minds that it might be related to the investigation. McGrew likened it to other minor changes that Boeing had made to the 737 without any need to notify the safety board. "I must tell you that we probably won't tell you when we change the brand of tires that we start putting on airplanes, either."

Hall was miffed. He told McGrew that the Roll Team looked like a parallel investigation without the NTSB, That could damage the integrity of the safety board's effort, Hall said. The board should be notified of any work that might be related. "It's like the Holiday Inn. The best surprise is no surprise." McGrew took his licking on the stand without complaining further, but again he felt he was a victim of Hall's grandstanding. The NTSB staff not only knew about the Roll Team, McGrew said later, but there was an NTSB person on the team.

Still, McGrew faulted himself for not doing a better job of communicating with the NTSB, informing the top officials ahead of time. He was a great engineer, but he realized he still had a lot to learn about politics.

In an interview a month later, Hall was still complaining about Boeing's behavior. "I think somebody [at Boeing] went brain-dead, to think they'll put together a team to look at rolls, but it had nothing to do with this investigation."

Hall had promised Haueter and Phillips that he would let them keep the investigation going as long as they needed, until they had exhausted everything. But he was growing pessimistic. "It looks like it will be very difficult" to solve the mystery, he said. "I've always said that good luck comes only after hard work. We've certainly worked hard enough. Maybe that will help us."

When the rickety 737 landed at Boeing Field, the pilots were relieved. They said it had been a white-knuckle flight. The old bird just didn't have much life left in her.

The plane, Ship 213, was what USAir called a "runout." It had flown for twenty-seven years, but the airline found it was no longer economical to keep it. In contrast to the newer plane used in the Atlantic City flight tests, USAir squeezed every last mile out of Ship 213 and then donated it to the Museum of Flight in Seattle. The wings and tail were going to get chopped off so the fuselage could be used as a theater. John Little, the museum's assistant security manager, greeted the pilots as they got off the plane and peeked inside the cockpit. It must have been a scary flight, he decided. There was a Bible on top of the instrument panel.

A USAir maintenance official had realized that the plane would be ideal for a series of destructive tests of some of the most bizarre theories—that a rudder cable had snapped in flight or that a fat guy had stepped through the floor onto the cable. No one had pushed too hard for them. They were the kind of tests that might bend an airframe, so they couldn't be performed on a jet that would ever be flown again. But the retired plane gave them a perfect opportunity. The tests would allow the investigators to see the forest instead of the trees. They had run hundreds of experiments on individual components of the 737. Now they could see how everything on the plane worked together.

On the first day of the rudder tests, Haueter and Hall held a press conference at the Museum of Flight, which was adjacent to Boeing Field. The podium was set up directly beneath a green-and-white replica of the B&W, Boeing's first plane. Haueter played a video and explained how the tests would be conducted. Hall then announced that he was appointing a panel of "the greatest minds in hydraulics" to review the NTSB's work on the Hopewell and Colorado Springs crashes to see if there was anything else they should do. It was an extraordinary step that showed the NTSB was practically desperate.

Haueter didn't like the idea of an expert panel. Why did they need a bunch of so-called experts? Phillips knew more about the 737 rudder system than just about anybody on the planet. It was as though Hall was saying the NTSB wasn't smart enough to figure it out. But Phillips had actually been one of the people behind the idea. He and another NTSB engineer thought it would be helpful to have an impartial panel to bring new brainpower and validate the safety board's work. Maybe fresh eyes would see something that the safety board had not. Hall said his direction to the panel would be this: "If there's anything we missed, tell us."

As Hall spoke in the museum, the rudder tests on Ship 213 were under way down the street in a hangar where Boeing had once fixed B-52s. The pipsqueak 737 was parked between two gargantuan 777s, the shiny new Boeing aircraft that had just started flying. The tiny USAir plane had been opened up like a patient on the operating table. A tail section had been removed so Boeing technicians could attach wires to the PCU. Two Boeing vans were parked beside the plane. One was filled with computers and test equipment that would keep track of the results. The other van was like a life support system, pumping the 737's systems with false information so the old bird would think it was still flying. A gallon of Starbucks House Blend was parked beside the plane, to provide fuel for the investigators.

The fat guy theory had been discounted by just about everyone. McGrew joked that it was possible only if the fat guy wore high heels, which would have broken through the floor. But everyone had agreed to one last test. The plane's floorboards had been removed so the investigators could see the cables that ran beneath passengers' feet. Technicians climbed into the cargo bin, where they could look up through the cables into the passenger cabin. They hooked a ratchet onto the rudder cable that would simulate the weight of the fat guy. By cranking on the ratchet, they could add weight in 50-pound increments. The device had a weight gauge so they could watch as the guy went from skinny to obese. Other investigators stood outside the plane, watching to see if the rudder moved. If it slammed hardover, they would know the fat guy was more than just a joke.

"Ready?" someone asked.

"Okay," Phillips said.

The guy started at 50 pounds, more of a kindergartner than a fat guy. No movement of the rudder.

He got heavier, up to 100. Still no movement.

At 150 pounds, the rudder barely budged. At 200, there was a slight

151

movement, and at 250, a tiny bit more. The instruments in the van showed the rudder had moved only 3.2 degrees, an insignificant amount. (By contrast, the rudder on Flight 427 actually moved about 21 degrees.)

Later that day they tried the cable-cut test. Someone sat in the cockpit and moved the rudder pedals to make sure they worked properly, doing slow sweeps back and forth. Then the pilot took his feet off the pedals and a technician cut one of the cables with a big bolt cutter.

Twaang! The noise echoed like a gunshot through the huge hangar as the cable snapped and recoiled through the plane like a broken rubber band. Outside, Cox saw the rudder panel shudder, but it didn't turn. Technicians installed a new cable and cut it in a different spot. *Twaang!* But again, the rudder didn't turn.

They did more than one hundred other tests and the plane passed them all. The rudder system still seemed to be invincible. But Phillips remained surprisingly optimistic. He knew they were still a long way from finding the cause, but he was patient. Each test gave them new data and brought them another step closer.

Across the street in the M-Cab simulator, Boeing's lobbying campaign was under way.

M-Cab had become a tool of persuasion to show how Emmett and Germano could have prevented the crash. At McGrew's suggestion, Boeing engineers had installed a switch that allowed anyone sitting in the pilots' seats to take control of Flight 427 and, without much effort, to keep the plane from crashing. It was a roller coaster ride with a message. Point the nose down (Emmett or Germano had mistakenly tried to pull it up), twist the wheel, and you have saved 132 lives.

Boeing invited Hall to ride the latest simulation to see how easily the crash could have been avoided. If Hall, a country lawyer who was not a pilot, could save the plane, surely Emmett and Germano should have.

Boeing test pilot Michael Hewett, a balding former navy pilot who looked strong enough to bench-press a 737, led Hall across the ramp into the white cab and invited him to sit in the right pilot's seat. Hall would act as the first officer on Flight 427. The simulator would reenact a rudder hardover, and then Hall would have the opportunity to recover the plane to keep it from crashing.

Hewett believed that pilots had to assert themselves in the sky but that the crew of Flight 427 had failed to take charge. He said the cockpit tape showed that Captain Germano had "no command presence."

In the simulator, Hewett's point was simple: The pilots could have avoided the crash by easily turning the wheel completely to the right and pushing the stick forward to let the airplane gain a little speed. The 737 would have lost altitude, but everyone would still be alive. Hewett summed it up by saying that to pilots, "speed is life."

Hall buckled himself in and M-Cab started its imaginary flight, cruising along at 6,000 feet. Cox was also riding in the simulator, sitting in the observer seat just behind them. Hall had invited him along because he knew Hewett was going to do a hard sell and he wanted Cox to provide a counterpoint.

Hewett started by demonstrating the recovery himself. He said the standard reaction time for a pilot allows three seconds, so he would count off before he recovered. When the plane started to go nose down, as the phantom pilots pulled back on the stick, Hewett flipped the switch to take control of the plane and counted, "One-thousand-one, one-thousand-two, one-thousand-three." He then turned the wheel to the right, gained control of the plane, and leveled off. He had prevented the crash.

Now it was Hall's turn. When the rudder suddenly went hardover, Hall followed Hewett's instructions and quickly turned the wheel to the right. "Hold it! Hold it!" Hewett told him. The simulator started to plunge toward the ground, but he stopped the roll and brought the nose back up. "Ease it out," Hewett said. Hall had saved the plane.

"That's with no pulling or pushing at all," Hewett said. All he'd had to do was turn the wheel right and hold it there.

Hewett then mentioned one of the mysteries of the cockpit voice recorder—the fact that neither pilot had given any clues about what was happening to the rudder. If the PCU was jammed, "Wouldn't you be saying, 'It's jammed! Goddammit, it's jammed!'?"

They flew again and Hall again recovered. But Cox didn't like the way the session was going. It seemed to him that Hall was hearing only Boeing's side of the story. He told Hall that it was understandable that the pilots pulled back on the stick. If they looked out the window, all they saw was the ground looming closer. "The airplane is not responding the way they want it to," Cox argued. "The windscreen is full of the ground and it is understandable that they would try to reduce the number of variables that they are facing."

"But," said Hewitt, "anybody who has ever been trained in a jet knows, with the stickshaker going off, the only way to recover is to let up on the stick. His first reaction should have been to push up on the stick."

Hall tried again. "There's the rudder in full hard," Hewett said. "Right wheel! Right wheel!"

But Hall turned left. The plane crashed.

"I almost recovered," Hall said.

After the session, Hall said Hewett was too heavy-handed. The Boeing engineers had had nearly eighteen months to figure out how to recover. Emmett and Germano had had just ten seconds. "No one who was flying the plane had been trained what to do," Hall said.

19

BLAMING GOD

May 1965

FAA Offices

Renton, Washington

"This valve," said Boeing hydraulics engineer Ed Pfafman, "is what we consider our safety feature."

Another Boeing engineer boasted that the unique valve-within-a-valve and a separate backup unit would create three ways to move the rudder on the new 737, which he said was "above and beyond the call of duty." Some planes had only two.

But as the Boeing engineers explained the plans for the 737's rudder system, FAA officials were skeptical. They were worried that the valve-within-a-valve might not be sufficiently redundant.

"The thing that is disturbing me is that you have more eggs in one basket here," said FAA official Charlie Hawks."It does shake me a little bit," he added. "At this moment, at least, I'm still a little jumpy about it."

The valve had been invented by Robert R. Richolt, a young Boeing engineer who had designed sophisticated valves on the Lockheed Electra. He had been so successful that he had retired and was living on a yacht when he was hired by Boeing. The dual valve was first used on the rudder of the 707 and the elevators in the 727. Richolt's 1963 patent application for the 737 said the valve was "fail-safe" and that the device "provides an override feature in case the main slide becomes seized. . . . This eliminates the quick reaction time required of the pilot to prevent a crash."

Redundancy is a fundamental tenet in designing airplanes, a concept that engineers jokingly call the "belt and suspenders approach." If your belt fails, your pants are still held up by your suspenders.

On many transport jets, such as the older 727, there was redundancy on the tail itself. The rudder was split into an upper panel and a lower panel. If one malfunctioned, the other one could control the plane or be turned the opposite direction to neutralize the problem. But that wasn't the case on the 737, which, during Boeing's push for fewer parts and better reliability, was designed with one big rudder panel.

The 737 rudder system was unusual. On most big jets, rudders are powered by separate valves rather than by the valve-within-a-valve. Many planes, such as the newer 757 and 767, have a special feature called a "breakout." If two valves detect that the other one is jammed, they break it out of the system.

But despite their reservations in the mid-1960s, FAA officials ultimately decided that the unique 737 valve met federal standards. The rules back then were relatively vague — they said that manufacturers must protect against failures "unless they are extremely remote." That phrase was not defined, which gave Boeing leeway. The company convinced the FAA engineers that Richolt's invention complied.

The first 737 flew in early 1967 with Richolt's valve in its tail.

The FAA, an automatic party on every NTSB investigation, assigned more than a dozen employees to the Flight 427 probe. The team was led by Vikki Anderson, the former flight attendant, and included FAA test pilots, inspectors, engineers, and a doctor from the agency's Civil Aerospace Medical Institute. The team had representatives on each of the important NTSB groups. There was a lot at stake for the FAA because it had certified that the 737 was safe. If the plane got blamed for the crash, the FAA would also get blamed.

The investigation revealed two faces of the FAA: the aggressive midlevel employees who handle the day-to-day scrutiny of Boeing, and the cautious top-level managers who occasionally sound like Boeing cheerleaders.

Steve O'Neal was an FAA bureaucrat, but he looked more like a friendly bartender than a paper-pushing drone. He was a gentle man with thinning brown hair and a thick moustache, forty-two at the time of the crash, with two stepdaughters and two grandchildren. He was a dog lover (he had an old Samoyed named Casey who was losing her vision), and he was passionate about flying. He could never afford to do much piloting on his own, but his job as an FAA flight test engineer in Renton allowed him to take test flights several days each month. He usually rode in the cockpit to make sure the plane met each FAA requirement. He had flown thousands of times, but he still got a thrill when the pilots took the plane to the edge of its limits, stalling it at scary angles and then recovering.

O'Neal spent his days studying failure modes, the myriad ways in which an engine or a flight control could malfunction. He had to be sure that those failures were benign or extremely improbable or that there was an adequate backup system to provide redundancy. He was especially proud that he had caught a complicated failure mode on three Airbus models involving a hydraulic failure. Airbus changed its pilot procedures because of his discovery.

In the 427 investigation, O'Neal was assigned to the NTSB aircraft performance group, which analyzed the flight data recorder. He was an independent thinker and was not shy about making suggestions that might be unpopular with his bosses. In a meeting with FAA managers about a month after the crash, he suggested grounding the entire fleet of 737s. Two of the planes had crashed in similar circumstances, he said, and the safest approach was to keep them on the ground until the safety board figured out the cause.

The idea went nowhere. Everyone else in the room felt there was insufficient evidence for such a dramatic step, which would have catastrophic consequences for virtually every airline in the country. Realizing that his idea had no support, O'Neal backed down. He was a conscientious employee, but he was also realistic. He knew he didn't have proof that the 737 had a flaw. But he continued to have qualms about the plane and urged friends and relatives to fly other planes if possible. He worried that 737 pilots would not be able to recover if they had a rudder hardover.

The other face of the FAA was Thomas McSweeny, a top official in charge of airplane certification. McSweeny was a twenty-two-year veteran of the agency, having previously worked eight years for an aerospace company. He was not formally a member of the Flight 427 investigation, but as the FAA's chief regulator of airplanes, he was in charge of any NTSB safety fixes that grew out of the accident.

The difference between McSweeny and O'Neal was politics. O'Neal worked in a job that allowed him to be an aggressive regulator without much concern for the political realities of the FAA or the costs to the manufacturers. If he thought the FAA should do something, he would propose it, even if it was unlikely to get past the scrutiny of the FAA's economists and top officials. But McSweeny was a political animal who had to be realistic about what could be done.

McSweeny often came across as a cautious bureaucrat rather than an aggressive regulator. When he talked about safety proposals, whether they were for the 737 or some other airplane, he was likely to sound like a spokesperson for the manufacturer. One reason for that was the FAA's awkward position anytime someone questioned the safety of a plane. If the FAA failed to give a hearty endorsement that a plane was safe, the agency was essentially admitting it had done a lousy job. If the plane was unsafe, then the FAA must have screwed up by not spotting the problem first. McSweeny could have gotten around that easily enough if he'd just sounded as if he was tough on the companies he regulated, but in the view of NTSB officials, he often came across like a salesman for the companies. Even when the FAA mandated safety changes to the 737, he sounded like a booster of the plane.

"They are really not design defects," he said during a press conference to announce rule changes for the 737. "They are *improvements* to the design." Boeing engineers used the same language to explain any fixes to the 737, calling them "improvements to an already safe design."

Relations between McSweeny and the NTSB were strained. That was no surprise, since the NTSB was the aviation watchdog. It needed to uncover problems at the FAA to justify its existence. Safety board officials liked to grouse about the bureaucratic sluggishness of the FAA and how protective the agency could be of the companies it was supposed to regulate. "McSweeny drives us nuts," Haueter said. "It sounds like he's speaking for Boeing."

Likewise, many people at the FAA complained about the safety board's political grandstanding and its holier-than-thou approach. Relations were further strained because the safety board got mountains of favorable press coverage and the FAA often got pummeled.

McSweeny seemed to have a strong dislike for the NTSB. He occasionally asked his staff to find out the agency's next move so he could preempt the board with regulatory action. He told FAA staffers that one of his goals for the 737 Critical Design Review was to beat the NTSB to the punch. If everything worked out, McSweeny would get his report back long before the NTSB solved the case and the FAA would get the credit.

The review began two months after the Hopewell crash and took about six months. The result was an inch-thick report that made twenty-seven recommendations about the plane. It called for better pilot training, fixes to the 737's erratic yaw damper, and clarifications for vague language in the federal air regulations. But when FAA officials described the findings in an executive summary, they made it sound as though the report was a ringing endorsement of the plane. They stressed that their review found no serious problems with the 737. "No safety issue has been found that requires immediate corrective action," the executive summary said. The twenty-seven recommendations will "enhance an already safe design."

As Brett walked into the thirteenth-floor conference room on September 23, 1996, to give a deposition, he thought about snubbing USAir lawyer Ann P. Goodman by refusing to shake her hand. After all, she represented the airline that had killed Joan, so it seemed perfectly reasonable that he didn't have to be nice to her. Brett, wearing a tie that Joan had given him shortly before the crash, had come to the offices of McCullough, Campbell, and Lane in Chicago so the lawyers could determine how much Joan's life was worth. USAir had the right to ask probing questions about their marriage, their income, even their sex life, to determine how big an award he was entitled to receive. Brett resented the whole process. "They killed my wife," he said before the deposition, "and now they get to invade my privacy."

It had been nearly two years since the crash, and Brett was finally showing the first major signs of healing. He had stopped wearing his wedding band, sold his house in Lisle, and moved into a lakefront condo in Chicago. For the first year and a half after the crash, he had worn his wedding ring every day. But he was afraid people were beginning to think he was a weirdo who could not let go of the past. He tried putting it on the chain around his neck with Joan's battered engagement ring, but they kept clanking together. So he put the wedding band away and watched the tan line on his finger gradually fade.

He had done a few freelance jobs, writing ad copy to sell day planners, but it was not very inspiring work. He did not want to be hawking calendars the rest of his life. So he immersed himself in a plan to build a chain of restaurants. He knew that people wanted more than food when they dined out; they wanted an experience. A new chain called the Rainforest Cafe had taken an environmental theme and dazzled customers with life-size mechanical animals and a realistic jungle. He thought, *Why not do the same thing with dinosaurs?*

The appeal of dinosaurs was timeless. Every generation of kids seemed to be progressively more obsessed with them, yet no one had thought to franchise them. So Brett read everything he could about dinosaurs, the restaurant business, and the making of *Jurassic Park*. He called companies that made mechanical dinosaurs and asked them for pictures, videotapes, and prices. To avoid giving away his idea, he said he wanted to start "a small private collection of dinosaurs."

He knew it was a long shot, but it was the perfect time to try something new. He had no dependents, few obligations, and little to lose. He visited prehistoric exhibits and dinosaur companies in Utah, Arizona, Kentucky, and Michigan. Some of the mechanical beasts looked surprisingly realistic. Others looked old and ratty. Back home in Chicago, he scoured the Internet for information about theme parks and entertainment companies. He had no experience in the restaurant business, but he got help from his father, who was an executive in a company that ran school cafeterias.

Brett came up with a name—T. Rex's Dino World Cafe—and hired an artist to do a colorful logo of a tyrannosaurus with a mischievous grin. He also came up with a slogan ("The only restaurant where you're not at the top of the food chain") and designed a business plan and a presentation for investors. The centerpiece of each cafe would be a life-size T. Rex. Another area in the restaurant would have velociraptors, which were scary, long-legged dinosaurs that would periodically run out of the bushes and look as though they were about to attack customers at their tables. There also would be an area for families with small children that would have a tame brachiosaurus. The project consumed him. The more he worked, the less he thought of the crash.

In response to Brett's lawsuit, Boeing and USAir had blamed God. The airline said it had made all the necessary inspections of the plane and had followed federal rules. It said the crash "resulted from an Act of God, unavoidable accident, sudden emergency or conditions or occurrences for which USAir is not liable or responsible."

Boeing said essentially the same thing in its response. "Plaintiff's damages, if any, were directly and proximately caused by an act of God for which Boeing is not liable." It was boilerplate language used as a standard defense in many lawsuits, but when Brett heard about it, he was furious. How could they blame God for the crash? That made it sound like God *willed* the crash to happen. The companies seemed to be ducking any responsibility.

Now he was in a law office taking an oath to tell the truth about Joan. He decided to be respectful, so he shook Goodman's hand and sat down in a chair that looked out over the Chicago River. The court reporter sat to his left, while Goodman took a seat directly across from him. Mike Demetrio,

Brett's lawyer, sat at the far end of the table and appeared to be immersed in a magazine. But he was actually listening carefully to every word, ready to object at any moment.

"Please state your full name for the record," Goodman said.

"Brett Alan Van Bortel."

"Let the record reflect this is the discovery deposition of Brett Alan Van Bortel being taken pursuant to the notice for today's date," she said. "Mr. Van Bortel, my name is Ann Goodman. I'm one of the attorneys representing USAir in this matter."

Goodman was a petite, humorless woman with blond hair who was the Chicago-based lawyer for USAir. Dombroff, the airline's lead attorney, worked in Washington, so he had retained Goodman's firm to handle much of the work on the Cook County lawsuits. Reading from Dombroff's lengthy script of questions, Goodman robotically asked about Brett's homes and mortgages. Then she abruptly switched gears and asked about drugs. Each question elicited just a fragment of information, so it took ten or twenty questions to get to anything of substance.

"Are you currently taking any medication?"

"Yes, I am."

"Have you taken any medication today?"

"No, I have not."

"Have you taken any medication in the last twenty-four hours?"

"No, I have not."

"What medication are you currently taking?"

"On an irregular basis when I travel, Xanax. It's an anti-anxiety drug for air travel, I guess."

"And you only take it for air travel?"

"Before and during flights. That's about it."

Goodman asked lots of questions about Brett's employment, from the days when he operated a ski lift in Vail through his writing job for the company that published the *Official Airline Guide*. Then she began asking about Joan.

"Did you marry a woman by the name of Joan Lahart, correct?"

"Joan Elizabeth Lahart."

"How did you happen to meet?"

"A friend of a friend, very casual, distant type of acquaintance. I had only met Joan briefly and it wasn't until later on that we started dating."

"What was it that attracted you to Joan?"

"I don't know that I could put that in a nutshell," Brett said. "It would be a combination of many things, but I thought she was a very beautiful woman and a very strong-willed and motivated woman."

Goodman seemed to jump randomly from topic to topic. She asked about Brett's and Joan's hobbies and interests, whether they went to college or professional football games, then about the names and ages of each of Joan's brothers, then how much they had paid for the house on Riedy Road, and then about the night of the crash.

"Was Joan required to make a business trip for Akzo Nobel on September 8, 1994?"

"I guess I would quibble with *required*," Brett said. "I don't know if they *made* her, but she went on her trip for Akzo, yes."

Goodman asked Brett to describe the last time he had seen Joan and how he found out about the crash. Demetrio had heard the same painful questions at several depositions with his clients, and he was mystified as to why the USAir lawyers persisted in asking them.

"Did you see her on September 8 before she took off for the airport?"

"Yes."

"When did you see her on September 8th?"

"It was our ritual for me to drop her off at the train station in Lisle for her 6:20 train, which I did that morning."

"What did you say to her, and she say to you."

"I think I just said, 'Love you and good-bye.'"

"Did you talk to her during the course of the day?"

He nodded.

"Did she say anything to you at the train station?"

"She just, I think, repeated that back to me. We knew we'd speak on the phone later that day."

Goodman asked about every tragic detail—what they said when they spoke that afternoon, how he heard about the crash, when he called USAir, when he got confirmation that she had been on the plane, and when he received Joan's remains. Then she asked about Brett's visits to see a psychologist.

"Did Dr. Pimental help you?"

"In some ways, but ultimately, no."

"How was she able to help you?"

"I would say helping me understand myself better and my reaction to it, but I ultimately came to the conclusion that no one but God was ever going to change what happened, and talking about it would not help me."

Back to mortgage questions. Goodman wanted to know how much Brett had paid for his new condo on Lake Shore Drive, how big a mortgage he had, what his interest rate was. And then she abruptly shifted gears again.

"How would you describe your marriage to her?"

"Excellent. We got along like best friends. I was very fortunate. I don't know why or how it happened, but I was one of the people that had one of the very good ones. I was very lucky."

"Did you plan to have any children."

"Yes."

"Had you made any attempts to start a family?"

"No."

"Were you waiting for a certain period of time?"

"Yes."

"How long had you planned on waiting?"

"About another year."

"How would you describe your physical relationship with Joan?"

"Very good. I don't know, I guess I'm uncomfortable describing it. It was very good."

"You had normal sexual relations with her?"

"Yes."

"On a regular basis?"

"Yes."

"What was the frequency?"

Brett looked at Demetrio to see if he would object. When Demetrio did not, Brett looked back at Goodman, pitying her for having to ask such a question. *She probably goes home and hates herself,* Brett thought.

Brett answered the question.

Goodman asked who did the cleaning, took out the garbage, did the laundry, shoveled snow off the driveway, did the grocery shopping, the cooking, and the dishes. Who paid the bills? Did he have his bank statements from 1994? How much was the electric bill? The gas bill? How was Joan's health? Did she smoke? How much did she drink? How much did she weigh? Was she ever convicted of a felony or a misdemeanor?

Brett answered all her questions.

"Are you currently engaged?"

"No."

"Have you dated anyone recently?"

Demetrio had known that one was coming. "I'm going to object to that question for the same reasoning that I objected to it at all the other depositions, and based upon that particular question, I'd instruct him not to answer," he said.

"I will state what my answer to your objection is," Goodman replied, "that it goes to the discovery process."

"Great," Demetrio said, still unconvinced. "I understand, just like I did all the other times."

Goodman abandoned the question and went to the next one. "Do you have any plans to remarry?"

"No," Brett said.

20

EASTWIND

As Eastwind Airlines Flight 517 neared Richmond, Virginia, the night of June 9, 1996, Captain Brian Bishop felt a bump from the back of the plane. "Did you feel that?" he asked the first officer.

Before Bishop got an answer, the plane's nose veered right and the right wing dipped toward the ground. Back in the passenger cabin, a flight attendant was tossed into a row of seats. Bishop stomped on the left rudder pedal, but it felt stiff. He turned the wheel to the left and added power to the right engine. That stopped the plane from rolling, but he could not get the wings level. He pressed on the left rudder pedal with all his weight, but could not get it down. The plane was stable, but was flying with the right wing tipped precariously toward the ground. The 737 was heading straight for the lights of downtown Richmond.

Bishop glanced out the window, looking for an area with no lights. If he had to put the plane down, he wanted to do it away from homes and buildings. Suddenly the rudder seemed to release and the wings leveled off. Bishop told the first officer to start the emergency checklist for a sudden roll.

"Autopilot off," the first officer said.

"Off," Bishop said.

"Yaw damper off." One of them reached to the switch on the ceiling.

"Off," Bishop said.

But then there was another thump, and the right wing dipped again. It seemed as if the pilots were losing control.

"Declare an emergency," Bishop said. "Tell them we've got a flight control problem."

The first officer relayed that message to the Richmond tower. A controller gave them a new heading to the airport, but Bishop was having trouble turning the airplane. He had the wheel cranked to the left and was putting most of his weight on the left rudder pedal, but he could not get the wings level. He was worried that he might not make it to the airport.

Then the rudder seemed to release again, allowing Bishop to level the wings. His first officer told the controller they could make the turn toward the airport. They were about five or six miles away now. They hurriedly went through the landing checklist. Bishop knew that a plane was more vulnerable to rudder problems when it flew low and slow, so he told the first officer, "I'm going to stay high and fast." Bishop was afraid that if the strange problem happened again he would not be able to recover.

As they descended toward the runway, fire trucks were waiting with red lights flashing. The plane touched down and rolled almost to the end of the pavement. The first officer told controllers the plane was okay and asked that the fire trucks stay away so passengers would not be alarmed. As they taxied to the gate, Bishop realized that he was so scared his knees were shaking.

He picked up the microphone to make an announcement to the passengers, but then he wondered what he would say. Anything he said would just make it worse. He put the microphone back.

Haueter and Phillips heard about the incident the next day. At first they thought it was a minor malfunction unrelated to the USAir crash. But then they discovered that Bishop had reported earlier rudder problems with the same plane. They headed for Richmond.

Eastwind was a new airline based in Trenton, New Jersey. It had only two planes, both of them 737-200s formerly flown by USAir. The planes had been repainted with Eastwind's logo, a squiggly line along the windows and, on the tail, a bumblebee wearing sunglasses. The airline had dubbed itself "the Bee Line."

Three weeks before the Richmond incident, Bishop had felt a slight rudder kick in the same plane when he departed Trenton and leveled off at 10,000 feet. It felt like the copilot had tapped the rudder pedals. He circled back and landed. Mechanics inspected the power control unit and replaced a coupler for the yaw damper, a device that made tiny adjustments to the rudder to keep the plane flying straight. The 737 yaw damper had a reputation for frequent malfunctions, so that seemed to be a logical fix. (The yaw damper could move the rudder only 2 or 3 degrees, so Haueter and Phillips did not believe it could have caused the full 21-degree movement that led to the crash of Flight 427.) Bishop tested the plane the next day, and the rudder pedals felt fine. The plane went back in service.

Now it was on the ground again, being dissected by engineers from the NTSB, USAir, and Boeing. They discovered some curious things. When mechanics had hooked up the yaw damper, they rigged it wrong. Instead of being limited to 3 degrees either direction, the yaw damper could move the rudder 4.5 degrees right and 1.5 degrees left. (That had not been a problem on Flight 427, however. The PCU had been rigged correctly.)

The Eastwind flight data recorder was rushed back to the NTSB lab to see if the engineers could decipher what had happened. It showed the plane had some strange rudder movements. It had initially gone to 4.5 degrees, presumably because the device on the yaw damper had been rigged wrong. But there was a second movement when it went to about 7 degrees and stayed there for twelve to fourteen seconds. That big a movement could not be explained by any sort of misrigging. It looked like the rudder had gone hardover.

The thirty-nine-year-old Bishop was a wiry former commuter pilot who did not have the polish of the typical airline captain. He had stringy brown hair and was always dashing to airport smoking lounges for a quick cigarette before departure. He had had the misfortune of working for two airlines that went out of business, so he had driven an airport snowplow until he got the job with Eastwind. He had a gritty personality that Haueter liked. His experience in flying small commuter planes came in handy when the rudder kicked. He used asymmetric thrust—putting more power in one engine—to keep the plane flying straight. That was a common approach for a "throttle jockey" flying a small turboprop, but it was rare for a 737 captain.

Boeing officials said Bishop had overreacted. Sure, there had been a malfunction in the yaw damper, the company said, but pilots were notorious for exaggerating their accounts of a sudden roll. Boeing said there were problems with gyros feeding information to the Eastwind flight data recorder that raised doubts about whether the rudder had truly gone to 7 degrees.

Haueter thought that the incident gave him a unique opportunity to test an airplane that may have had a hardover. He knew Eastwind was crippled while the plane was grounded in Richmond—the plane was half of the airline's fleet—but he couldn't pass up the opportunity to try flight tests with a crew that might have encountered a hardover.

The parties agreed on two tests. One would measure how a 737 would react to a yaw damper problem. The other would see how a pilot would react to a sudden rudder movement. Bishop would be the guinea pig.

The day before the test, Michael Hewett, the Boeing test pilot, and several NTSB investigators arranged a conference call with Bishop to explain how the test would go. As usual, Hewett was brash—so brash that Haueter thought he was trying to bully Bishop into thinking he hadn't responded the right way.

"Stop it!" Haueter said. "You're trying to intimidate this guy."

Hewett said he was just trying to get Bishop to understand what had happened that night. Hewett seemed to have doubts about Bishop's competence. "When I put my wife and children on an airliner, I expect the people flying up front to be as good as I am."

Haueter was furious. He thought that Boeing was trying to influence the test.

Hewett was equally angry. He thought the NTSB was acting like the Gestapo, limiting what questions he could ask. He wasn't trying to influence Bishop; all he wanted was an accurate story from him.

The first test was done without Bishop. The Eastwind plane was configured exactly the way it had been when Bishop had his scary incident at 4,000 feet, with the same PCU, the same yaw damper. The only change was in the cabin, where seats had been removed so Boeing could install its flight-test computers. The mood was tense at a preflight briefing as Hewett and the NTSB went over the test plan. They didn't know if there was a gremlin in the tail, and no one could be sure that Hewett would be able to control the plane if something went haywire. The route took the plane out to a restricted military area over the Atlantic Ocean so they could try maneuvers away from a populated area. If the rudder went hardover, they wouldn't wipe out a neighborhood. But Hewett was not nervous at all. He thought the NTSB was overreacting by insisting that the plane fly over the ocean.

Once the plane climbed into the sky, Haueter figured it was safe. The critical time was takeoff and landing. When a plane was flying faster than 200 knots, pilots could easily recover from a hardover.

Hewett and an FAA test pilot flew out over the Atlantic and put the plane through the maneuvers, kicking the rudder right and left. The plane flew poorly because its controls seemed to be out of alignment, but the pilots found no problem with the rudder. They returned to Richmond.

Before Bishop got on the plane for the second flight, Haueter warned him about Hewett. "Look, the purpose of this is I want to know your perceptions. Don't let anybody talk you out of anything. Take a piece of paper so you can write something down immediately after it happens so it will be fresh. We want to know how similar this is to what you had the night of the ninth, as best as you can recall."

Bishop took the controls and flew the same as he had on June 9. Without warning, Hewett gave a quick hand signal. An FAA pilot behind them pushed a button that suddenly moved the rudder 4.5 degrees. The plane started to roll.

Bishop quickly stomped on the opposite rudder pedal, which stopped the roll and made the plane roll back toward wings level.

Was that what happened on June 9? Hewett asked.

No, Bishop said, the test was much slower. "This isn't even a tenth of what we felt that night."

"Well, it was dark out, you weren't expecting it," Hewett said. He seemed to be offering excuses to show that Bishop had exaggerated.

"This wasn't even close," Bishop said.

Haueter had studied dozens of suspicious 737 incidents since the crash and had found reasonable explanations for nearly every one. Many were yaw damper problems, some were autopilot malfunctions, and lots were encounters with wake turbulence. But the more he studied Eastwind, the more it matched Flight 427. He believed Bishop's account, which was corroborated by the first officer, that the rudder pedal would not move. The flight data recorder also verified their accounts and showed that the plane was cross-controlled, meaning that the rudder was turned at the same time the ailerons were going the other direction. The only problem in the pilots' stories was their confusion about the timing of the events, which was common for all pilots. Haueter concluded that the Eastwind incident was eerily similar to 427, with one important difference. The plane's speed was 250 knots, well above the crossover point, so Bishop was able to recover before it crashed.

On the two-year anniversary of the crash, by pure coincidence, John Cox took a USAir 737 to Pittsburgh. He was scheduled for his six-month training session at USAir's simulator center near the Pittsburgh airport. As the plane approached the city, Cox was sitting in Row 8 of coach, studying a pilot handbook about things that can go horribly wrong on a plane—engine fires, takeoff stalls, autopilot failures. The passengers sitting around him might have gotten heartburn if they had seen what he was reading, but to Cox it was like studying for final exams.

The USAir emergency checklist had been improved because of Flight 427. Now there was a procedure for "Uncommanded Yaw or Roll," which called for the pilot to turn off the autopilot, grab the wheel firmly, and turn off the yaw damper. The cover of the checklist also had changed, listing the general approach for pilots anytime a plane was in trouble:

> Maintain aircraft control
> Analyze the situation
> Identify the emergency
> Use the appropriate checklist

Cox wasn't nervous about his six-month checkup, but he wanted to make sure he aced it. He double-checked the list of maximum fuel temperatures and silently recited the engine fire procedure ("Auto-throttle off, throttle idle, start lever cutoff, fire handle pull. If light doesn't go out in thirty seconds, rotate handle"). As the flight attendants prepared the 737's cabin for arrival in Pittsburgh, he was so caught up in preparing for calamities that he didn't notice the time—7:03 P.M.

Two years to the minute since Flight 427 had crashed into the hill.

The next day in the simulator, he dealt with one crisis after another. He was climbing out of Philadelphia, leveling off at 10,000 feet when *bang!*— the No. 2 engine seized up. He went through the memory procedure for an engine fire. Once the situation was under control, he pulled out the checklist and started going through it with the first officer. An instructor pounded on the cockpit door, pretending to be a flight attendant.

"What is going on!? What was that bang?" the instructor shouted.

The weather in the simulated Philly was lousy, so Cox diverted the plane to Baltimore and landed with one engine. He then departed for Charlotte and found horrible weather there, too. He had to use the 737's auto-land system, which can land the plane in thick fog. More mishaps followed. The plane began to stall after takeoff. Another engine problem. And then his beeper went off.

He glanced down at the beeper and saw the number. It was LeGrow, the union's chairman for the Flight 427 investigation. At the next break Cox walked to the pay phone and called him.

"Hey, Herbie, what's going on?" Cox asked.

"John Boy, have you checked your Aspens?" LeGrow asked, referring to the ALPA voice mailbox system.

"No."

"You're going to like it," LeGrow said. "Kenny has the safety recommendations. They couldn't be any better. They are just what we wanted."

"Yes!" Cox said, punching his fist in the air like a pitcher who just struck out a .300 hitter.

He walked back to the simulator and told the instructor, "Four-twenty-seven is drawing to a close. And we're going to like the results."

Greg Phillips's recommendations for the 737 had come back to life because Laynor, the safety board official who had blocked them, had retired. They found a much more receptive audience in Bernard Loeb, the new head of aviation safety, and his deputy, Ron Schleede. That broke the logjam and meant that Haueter and Phillips could try again to get approval. The recommendations that were going to the board members were essentially the same list from Phillips's memo eighteen months earlier. They called for sweeping changes to the 737 and the way it was flown. They did not specify what Boeing should do—"We don't want to be junior engineers telling them how to do it," Haueter said—but they essentially meant that Boeing would have to install a limiter on the rudder or change the ailerons so pilots would have more roll control. The list also called for fixes to the yaw damper, a better method for pilots or mechanics to detect a jam in the valve, a new cockpit indicator to show pilots when the rudder moved, and new procedures telling pilots how to respond to a hardover.

Cox was ecstatic when he returned to St. Petersburg the next day, calling the recommendations a watershed event that would prevent future crashes. "I'm so happy, I am doing double back flips," he exulted. "We've now got the mechanism in these recommendations to fix the airplane."

McGrew was feeling burned out. He was overdue for a vacation and felt drained. His son said he looked like he had aged five years since the crash. It had been only two. He had lost enthusiasm for the investigation and thought that others at Boeing were in the same rut. In the first two years there had been a high level of energy throughout the company to solve the mystery, but nearly everyone had since moved on to other projects and now when they worked on 427 it was hard to get them fired up again.

To make matters worse, he had been removed from his job as the 737 chief engineer and given a different position, overseeing new Boeing models. It was a lateral move, but he did not want to go. His bosses didn't mention burnout as a reason for his reassignment, but McGrew figured that was one of the factors. He stayed involved in the Flight 427 effort, but he found less time each week to work on the investigation.

He was frustrated at the NTSB's lack of effort in studying the pilots. Sure, the human factors team had pursued several leads that Boeing wanted, but McGrew had heard that Brenner, the NTSB human factors investigator, believed the pilots had their feet on the floor and never knew when the rudder went in. How could Brenner say such a thing? There was no conclusive proof of that, and in McGrew's view, there was evidence to the contrary. McGrew and most people at Boeing still believed that Emmett or Germano (most likely Emmett, since he was the flying pilot) had mistakenly slammed his foot on the pedal and then pulled back on the control column, stalling the airplane and causing the crash.

McGrew and other Boeing officials had been traveling the world to reassure airlines about the safety of the plane. They were under tremendous scrutiny. A series of stories in the *Seattle Times* said Boeing had not responded to the rudder problems, despite many incidents. McGrew found that allegation preposterous. He said the company had thoroughly investigated the incidents and had found no systemic problem.

He said he felt no pressure from Boeing management to defend the plane and had done so only because there was no evidence that the rudder system had malfunctioned. Likewise, he said, the company's costs for lawsuits had no effect on what he did. "If it's a ton of money, that's too bad," he said one day while driving up Interstate 405 to a meeting. "If there's something wrong, you've got to fix it." He was convinced that Emmett and Germano just got into a situation that was over their heads.

McGrew and other officials thought it was time to throw the equivalent of a Hail Mary pass. They would go over Haueter's head directly to the board members, the five political appointees who would vote on the probable cause. Rick Howes, the Flight 427 coordinator in Boeing's air safety investigation office, made a courtesy call to Haueter. Howes said Boeing was going to be "aggressive" in informing the board.

The result was a spiral-bound booklet called the "Boeing Contribution to the USAir Flight 427 Accident Investigation Board." It was the classic Boeing approach—slick color renderings of what Germano saw from the cockpit and a view from behind the plane, matched with Boeing's analysis. The renderings were the same ones that Haueter had seen on the posters in the conference room a year and a half earlier, but this time they had Boeing's comments on why the plane was innocent:

> The rudder system on the Boeing 737 airplane has been operated successfully for 73 million flight hours. In all of these hours of service history, there has been no known occasion when there was a full uncommanded rudder input. Nor has there been any known occasion when a rudder malfunction produced an event that was not controllable by the flight crew. Most important, the extensive investigation conducted to date into the rudder system used on USAir Flight 427 confirms that this system was fully operational during the upset.

Boeing was careful not to criticize the pilots too much. The booklet said it was understandable that they had made a mistake after they were startled by the wake turbulence: "It is known that pilots respond to roll upsets by using rudder. It is also well documented that rudder inputs once made can be forgotten or ignored and maintained for the remainder of the flight." The booklet said Emmett and Germano were surprised when they were jostled by the wake turbulence. They tried to respond to the roll, but overcorrected with the left rudder pedal.

"In all likelihood, the crew became absorbed in making other control inputs as the upset sequence developed, and simply failed to perceive that a full rudder input had been made." The booklet quoted from a USAir document that said when pilots respond to upsets, "our biggest problem has been stepping on the *wrong* rudder!!" The booklet said there was no evidence of any failure of the rudder PCU and that Flight 427 would have been recoverable if the pilots had not pulled back on the stick. It ended with this:

> Boeing believes there are persuasive reasons to support a conclusion that the USAir Flight 427 accident was caused by an unexpected encounter with wake turbulence, rudder commands by the crew and a failure to apply correct recovery techniques.

Haueter believed that the "Boeing Contribution" was designed to kill Phillips's safety recommendations. The booklet didn't mention them, but it arrived the week before the board was scheduled to discuss them and it clearly tried to deflect attention away from the plane.

But the booklet ended up having no effect on the recommendations. The safety board delayed them by two weeks so that Bob Francis, the board member working on the TWA 800 crash, could have more time to review them, but the safety fixes were approved unanimously with no significant changes.

The recommendations were sent to David Hinson, the FAA administrator, in an exhaustive twenty-six-page letter that cited the Eastwind incident and the crashes of USAir 427 in Pittsburgh and United 585 in Colorado Springs. The letter said the investigation of the USAir crash had not been completed, but that the NTSB had found 737 safety problems "that need to be addressed."

McSweeny, the FAA aircraft certification chief, was his usual defensive self in responding to the recommendations. He downplayed their importance, saying that the FAA had been addressing many of the same issues with previous airworthiness directives that grew out of the Critical Design Review. He said they would consider the NTSB's recommendations but that "you want to be very careful when you change [the 737]. You need to make sure that your tinkering does not cause a problem." Once again, McSweeny sounded as though he was protecting Boeing.

21

THERMAL SHOCK

1966

Bendix Electrodynamics

North Hollywood, California

A hydraulic valve had to pass a battery of tests to get accepted by Boeing. One test shook it violently, like a can of house paint in a mixer. Another test moved the valve back and forth 5 million times. The most brutal test froze the valve to minus 40 degrees Fahrenheit and injected it with hot hydraulic fluid. That represented the worst imaginable condition—an overheated hydraulic pump when the plane was in frigid air at 35,000 feet. Hot fluid would shoot into the frozen valve, causing it suddenly to expand. The test was called thermal shock.

Boeing did not manufacture its own valves, just as it didn't build most of the parts for its planes. Instead, it relied on hundreds of suppliers such as Bendix. The company did not make the unique valve for the 737's rudder, but it was bidding to make a similar one for Boeing's giant new plane, the 747. Bendix engineers built a prototype of the valve to undergo the standard battery of tests—the paint shaker, the marathon, and thermal shock.

The tests were held in a gray stucco building in an industrial section of North Hollywood, not far from the Burbank airport. The lab, which took up most of the first floor, was filled with a thick, oily smell from all the hydraulic fluid. The room was a veritable torture chamber for a hydraulic valve. The lab even had special steel containers called crash boxes that were used the first time a valve was pressurized, in case it exploded.

Upstairs was a man named Ralph Vick, an engineer who worked on some of the company's most important projects. Vick was not directly involved in the bid for the 747 valve, but he kept close tabs on the tests because he—like everyone else in the company—desperately wanted to win the big Boeing contract.

The torture tests on the 747 valve were no different from hundreds of others performed in the Bendix lab that year. The technicians placed the valve in a tiny freezer and hooked up the hydraulic lines. Once the valve had cooled to sub-zero, they flipped a switch and heard the steady whine of the hydraulic pumps. They moved the valve back and forth, as if a pilot were stepping on the pedals. Then someone flipped another switch, and piping hot fluid shot inside. Usually the valve kept moving. But this one strained and then stuck for a few seconds.

It had failed the test.

When Vick heard about the results, he knew it was a setback but not a catastrophe. The valve was an amazingly tight device, with only a few millionths of an inch between each slide, so a very tiny design error could cause a jam. The Bendix engineers went back to their drawing boards and redesigned the tolerances. The new valve passed without problems.

May 1996

L'Enfant Plaza Hotel

Washington, D.C.

Vick unpacked his suitcase in his hotel room and sat down at the desk with a legal pad. He had come to Washington for the first meeting of the "Greatest Minds in Hydraulics" to review Haueter and Phillips's work. The goal of the panel was to look for new tests that the safety board should try. At sixty-seven, Vick was a quiet, serious man, a good choice for the group because he had designed dozens of valves and had been awarded twenty-five patents. He was quite familiar with the unique valve-within-a-valve used for the 737 rudder.

Sitting in his hotel room, he recalled the Bendix test thirty years earlier when hot fluid hit cold metal and the prototype valve stuck for a few seconds. That jam turned out to be no big deal—a redesign took care of the problem. But he wondered if the rudder valve on the USAir plane had stuck the same way. He sketched a brief outline of the test on a piece of paper and gave it to Phillips the next day.

"I think we should look at this," Vick said. "It may be something."

The NTSB had not done a thermal shock test on the 427 valve because there had been no comments on the cockpit tape about a hydraulic problem. If one of the pumps had broken, it would have triggered a warning light in the cockpit and the pilots would likely have mentioned it. But Phillips agreed to try the test. He was open to any suggestion.

The power control unit from the USAir crash would be frozen to 40 degrees below zero, similar to the outside air temperatures at 30,000 feet, and then it would be pumped with hot hydraulic fluid.

No one expected a breakthrough. The 737 valve had passed its own thermal shock test when it was certified in the 1960s. Besides, the temperature range was far more extreme than anything the PCU encountered in real life. Boeing officials viewed the test as a waste of time. McGrew said a 737 would encounter thermal shock conditions only if it flew to the moon.

On August 26, the Greatest Minds in Hydraulics and Phillips's systems group gathered at Canyon Engineering, a tiny hydraulics company in an industrial park in Valencia, California. Cox thought the place looked more like a garage than a modern test facility. They had chosen Canyon because the chairman of the expert panel worked there, but the company did not have the sophisticated test equipment that Boeing and Parker did. Phillips brought the PCU in a sturdy navy blue chest, like a violinist carrying his prized Stradivarius. He took the 60-pound case to his hotel room each night to make sure that no one could tamper with the device.

At Canyon the PCU was placed in a big white Coleman cooler, the same kind you would take on a picnic. It looked like an amateur setup. Holes were cut in the cooler for pipes and tubes and then sealed with gray duct tape. When someone opened the cooler, frost formed on the PCU, making it look like a giant Popsicle. Cox and several others in the room said they were concerned that the temperatures were not controlled closely enough to produce legitimate results. But they forged ahead with the tests to see what would happen.

The group tested two PCUS—a new one straight from the factory and the one from the crash. To make sure that the hydraulic fluid was similar to Flight 427's, they used fluid drained from other 737s. They used a pneumatic cylinder to act like the pilot's feet, pushing the valve back and forth. The room filled with a steady rhythm of clicks and hisses as the cylinder moved the valve left and right.

Click, hiss, click, hiss. They put the factory PCU through its calisthenics at room temperature, testing it fifty times. It responded normally. They let gaseous nitrogen into the cooler and watched the temperature gauges plummet to minus 30 or 40 degrees, to simulate the cold air at 30,000 feet. *Click, hiss, click, hiss.* No problems. Finally, they tried two tests to simulate an overheated hydraulic pump, heating the fluid to 170 degrees. *Click, hiss, click, hiss.* The hot fluid hit the cold valve, but there were no problems. The factory PCU worked great.

They removed it from the Coleman cooler and installed the PCU from Flight 427. *Click, hiss, click, hiss.* No problems at room temperature. *Click, hiss, click, hiss.* The frigid unit was blasted with hot fluid, but it still worked fine. It was their last day in Valencia, and the tests were going so smoothly that several people started to pack up and say good-bye. Another theory had been ruled out. It was time to move on.

They had reached the most extreme condition. The PCU was depressurized, frozen with the nitrogen gas, and then injected with piping-hot fluid. They made the test especially severe by performing it with the A and B hydraulic systems separately, so the hot fluid would not be diluted by cooler fluid from the other system.

The hot fluid hit the cold valve. *Click, hiss, click, hiss, click, hiss, click, hisssssssssssssssssss.*

The hissing changed pitch. The valve had jammed.

"It didn't come back," said someone in the room.

"That's interesting," said someone else. "Reeeeaaalllll interesting."

A second later, the arm went back to neutral and began cycling again. *Click, hiss, click, hiss.*

They stopped the test and talked about what had happened. Did they have a breakthrough? Nobody could be sure. The test conditions were so poorly controlled that any result was questionable. A computer operator who had been collecting test data had mistakenly deleted everything, so they had little evidence of what they had seen. Everyone agreed to try it again.

Click, hiss, click, hisssssssss. Click, hissssssssss. The valve was moving slower than it was supposed to. *Click, hissssssssssssssssssssssss.* It stuck again.

They agreed that the test should be done again in a more controlled setting. The Boeing team criticized the tests, saying they were too extreme and that the valve could have been damaged. So the next morning, Phillips woke up at 4 A.M. and drove to Parker-Hannifin so they could perform a test to make sure the valve was okay.

The test was crucial. When they had first examined the valve after the crash, they had not found any scratches inside it. If they found scratches now, it would prove that a jam left a scratch, which would indicate there *had not* been a jam on Flight 427. Also, a scratch would mean that the valve had been altered since the crash, which would rule out any further tests. The whole theory about a valve malfunction would go down the drain.

The Parker technicians took the valve apart, measuring and documenting each piece. They put them under a microscope, examining each surface for scratches or scrapes. They found none and no evidence of a jam. Phillips breathed a sigh of relief.

They had proved that the valve could jam—and leave no evidence behind.

Six weeks later, Phillips's group reconvened in a Boeing laboratory in Seattle. This time, instead of testing the PCU in the Coleman cooler, they used a specially designed foam box with a window on top. The box's cooling system was more powerful and precise, with temperatures closely monitored by a computer.

They ran through the same tests they had done in Valencia, starting with the factory PCU at room temperature and then trying a variety of thermal shocks. There were no clicks and hisses this time because the pneumatic system had been replaced by a hydraulic one. In some tests, the technicians just pulled on a lever to move the valve. Once again, the factory PCU passed every test.

The technicians removed it and replaced it with the PCU from Flight 427. It passed the first tests with no trouble. Then came Condition G, a repeat of the most extreme test in the Coleman cooler. They removed hydraulic pressure from the PCU and let it soak in the cold air until it reached minus 40 degrees. The system A hydraulic fluid was heated to 170 and shot directly into the PCU. The technician moving the valve back and forth felt it slow down. He didn't notice it bind, but a computer showed it had jammed momentarily. They repeated the action, and the technician felt the lever kick back when he tried to move it to the right. He tried again, felt it stick to the left and then jam.

Once again they had shown that the 427 valve was unique. It jammed when the factory unit did not.

Yet Boeing was right. The extreme temperature range necessary for a thermal shock just didn't happen in real life. And there was no proof that it had happened on the USAir plane.

Despite their skepticism, Boeing engineers said they would examine the charts from the tests for anything unusual. They might learn how the valve and rudder reacted to a jam.

A few days later, in a building overlooking Paine Field in Everett, a young Boeing engineer named Ed Kikta sat at his desk, reviewing the charts. He could see the test data on his computer screen, but he liked to print the results so he could study them more closely. The charts showed the flow of hydraulic fluid during each test, higher when it was pushing the rudder and down to zero when it was not. Kikta expected that when the outer valve jammed during the thermal shock, the inner valve would compensate and send an equal amount of fluid in the opposite direction, which would keep the rudder at neutral. That was the great safety feature of the 737 valve. It could compensate for a jam.

But as Kikta studied the squiggly lines for the return flow, he saw dips that were not supposed to be there. When he matched them to another graph showing the force on the levers inside the PCU, he made an alarming discovery. When the outer valve had jammed, the inner valve had moved too far to compensate. That meant the rudder would not have returned to neutral, the way it was supposed to.

The rudder would have reversed.

That could be catastrophic. A pilot would push on the left pedal, expecting the rudder to go left, but it would go right.

To make sure he hadn't made a mistake, Kikta showed the results to the other engineers in the room. They agreed with his interpretation. It appeared that the valve had reversed. Kikta looked up and saw that his boss, Jim Draxler, was putting his coat on, getting ready to leave. Kikta stopped him.

"I think I've found something in the data," Kikta said. "We might have a problem here." Draxler took his coat off, set down his briefcase, and listened to what Kikta had to say. The consequences of his discovery were enormous. If he was right, it meant the PCU was not performing the way Boeing had promised. The valve-within-a-valve was supposed to provide redundancy if one slide jammed. But this meant a single jam could cripple a plane.

The next morning Draxler convened a group that he called his grizzled veterans, engineers who had lots of experience with flight controls. Kikta explained his findings and showed them the charts. Draxler went around the room, asking each one about the significance of Kikta's discovery. They were unanimous: It was a serious problem that needed to be fixed quickly.

Boeing sprang into action. The company ordered Parker Hannifin to run its own tests to check Kikta's conclusions. Parker engineers confirmed the results and discovered that when they jammed the outer valve, the levers in the PCU appeared to flex slightly, which allowed the inner valve to line up with the wrong holes.

Boeing was notorious for being the slow-moving "Lazy B," but not this time. Fear was a powerful motivator. Engineers usually needed weeks to get an airplane for a test, but now they got one off the assembly line in just twenty-four hours. The plane landed at Boeing Field and was pulled into the B-52 hangar where the fat guy tests had been held. As a cold downpour fell outside on the night of October 29, 1996, the 737 was rigged with the special device that Parker had built to simulate the jam. Hewett, the Boeing test pilot, climbed into the cockpit while Kikta stood on a platform on the tail of the plane, watching the rudder and the PCU. Hewett pushed on the pedals, moving the rudder from side to side. The first two tests went smoothly, and the rudder operated as intended.

Then came a more rigorous test. Hewett slowly stepped on the left pedal and the rudder moved properly. He then jammed his foot on the right pedal as hard as he could. It kicked back with tremendous force.

The rudder swung the wrong direction.

Further tests showed that the likelihood of the rudder reversing depended on where the outer slide jammed. If it jammed closer to its neutral position, the rudder was less likely to reverse. But if it jammed when it was farther from neutral, a reversal was more certain. It was about midnight now and everyone was exhausted. They all drove home worrying about what they should do to fix the plane. "Everyone was concerned," Draxler recalled later. "We didn't know what it meant, how it all fit together."

The next day, Boeing notified the FAA that the company had found a problem with the rudder PCU but wanted twenty-four hours to figure out how to deal with the problem. The FAA agreed.

Intense meetings went on all day in Renton and Everett as the Boeing engineers discussed how to respond. They broke into two teams, one to come up with a plan for how pilots could detect and respond to a jam and another to look at long-term design changes to the PCU. They worked into the night. By 11 P.M., they got approval from senior management for a pilot test and some short- and long-term changes to the PCU.

The next day, Halloween, about ten Boeing officials drove to the FAA office in Renton, a big mirrored cube of a building beside Interstate 405. They weren't sure what the FAA would do. Would the agency want to ground the airplane? The PCU no longer protected against jams the way it was supposed to, which could mean that the plane no longer met certification standards.

About twenty-five Boeing and FAA officials gathered in a conference room. McGrew took a seat just beneath a smiling photo of FAA administrator David Hinson. In the corner were a TV and a VCR, with the clock display flashing "12:00" throughout the meeting. Draxler began by explaining what they had found in the tests, with Hewett frequently interrupting to give his perspective. It took a unique kind of windup to trigger the phenomenon, they said. You had to press on one pedal and then stomp hard on the other to make the primary slide line up with the wrong holes and cause the reversal.

Did this match what had happened to the USAir plane? an FAA official asked.

The Boeing engineers said all they knew from the test was that if you jam the outer slide, you could get a reversal. Jams were extremely unlikely because of the many filters in the hydraulic system, which removed particles before they caused problems. In thirty years and more than 50 million flights, there had been only seven confirmed jams. None had resulted in an accident or injury. And there was no evidence that one had occurred on the USAir plane.

Another FAA official pointed out that the new evidence seemed to counter Boeing's claims that the pilots had caused the crash.

McGrew spoke up. "We've received a lot of public criticism about hiding things and not wanting to spend a lot of money," he said. "But I frankly don't care [what it costs]. If there is something wrong with the airplane, I want to fix it."

Steve O'Neal, the FAA flight-test engineer, was impressed by McGrew's comments. Until that point, O'Neal had felt McGrew was overprotective of the airplane. But McGrew now seemed sincere in saying that he was open to anything.

The meeting ended. Boeing said it would issue an alert service bulletin to warn airlines about the condition. The bulletin would require mechanics to perform a test every 250 hours, stomping on the pedals to check for jams. The FAA planned to issue an emergency airworthiness directive that mandated the tests. Boeing also said it would develop a long-term plan to redesign the valve to prevent a reversal. That fix was likely to take several years.

These directives were more symbolism than real action, designed to re-assure the public that the FAA was taking action. The engineers knew the tests would not be very effective. They would catch a jam if it occurred at the pre-cise moment of the test, but a jam could still occur at any time.

Despite the seriousness of Boeing's discovery, FAA officials say they did not give serious thought to grounding the 737 fleet. The plane had a good safety record, they said, and a jam was still considered highly unlikely. Even O'Neal, who had wanted to ground the fleet two years earlier, agreed with his bosses this time.

While the Boeing-FAA meeting was going on in Renton, Haueter and Phillips were 2,000 miles away in Pittsburgh, unaware of the dramatic developments. They had returned to the Holiday Inn near the Pittsburgh airport to meet with all the parties. Haueter and each of his group leaders gave updates on the investigation. Phillips reviewed the results of the thermal shock tests (without knowing of Boeing's finding) and discussed what work still needed to be done. Rick Howes, the Boeing coordinator for the investigation, sat through the all-day meeting without saying a word about the company's big discovery.

There was the usual sparring between ALPA and Boeing—this time over Boeing's latest estimates of the rudder movement—and then the meeting ended uneventfully. Haueter put a slide on the overhead projector with a comment that an English scientist once made to Charles Lindbergh. Haueter said it had become the slogan of the 427 investigation: "Everything that happens was once infinitely improbable. Therefore, nothing that happens should be surprising."

People chuckled and the meeting broke up. Haueter, Phillips, and Tom Jacky, the NTSB performance chairman, took a flight back to Washington. As they got off the plane at National Airport and walked toward the subway station, Haueter's beeper went off. The NTSB had a new pager system that could transmit words as well as phone numbers.

Haueter glanced down at it. MAJOR FINDING REL TO PIT / DEFECT FOUND ON SERVO VALVE, the pager said.

"This is a joke," he said. "This isn't real. Some jerk has figured out our paging system." They went their separate ways and headed home.

The message had come from Schleede, Haueter's boss, who had been working late in the NTSB office when McGrew and John Purvis, the head of Boeing's accident investigations, called to tell him about the finding. Schleede transmitted the message to Haueter and then walked downstairs to the bar at the L'Enfant Plaza Hotel, where Hall was having a drink.

"Jim," Schleede said, "I think we've got it."

The next day, the FAA briefed Haueter, Phillips, and other NTSB officials about the finding. Haueter realized that it was a major piece of his puzzle.

"This isn't the way I thought it would end," he told Phillips as they walked back after the meeting. "I expected it was going to be a fight all the way to the end, putting all these little pieces together, with people saying we wouldn't have enough evidence. And all of a sudden here is something no one expected."

That day, Boeing sent a telex message to every airline in the world that flew 737s:

Alert Alert Alert Alert Alert Alert Alert Alert
Boeing Alert Service Bulletin 737-27A1202 dated November 1, 1996
The dual servo valve is designed to overcome the effects of a jammed primary or secondary slide. Although there has never been a report of a secondary slide jam, tests just completed at Boeing have shown that, under certain conditions, some jams of the secondary slide can result in anomalous rudder motion.

Anomalous rudder motion. It was a Boeing euphemism for a catastrophic situation—a rudder jam and a reversal.

Cox heard rumblings about the discovery on Halloween night but didn't hear the news until the morning of November 1, when the alert was issued. He had spent an extra day in Pittsburgh and was summoned to the office of USAir vice president/flight operations William Barr. A group of pilots and safety officials were meeting to discuss the service bulletin and how USAir should respond. There were major implications for USAir because it had the third-largest fleet of 737s in the world.

Barr asked Cox point-blank, "Is the airplane safe?"

"Yes," Cox said. He was convinced that a jam was still highly unlikely and that, even if one occurred, pilots could recover. USAir had been the first airline to raise its minimum speed above the crossover point, so USAir pilots had an extra cushion of safety. And the airline's pilots were already doing a rigorous rudder check, which meant they were effectively conducting the test before every flight.

That afternoon Boeing issued a carefully worded press release that downplayed the seriousness of the discovery:

> Boeing recently discovered that, under certain conditions, a jam of the PCU's secondary slide could possibly interfere with the intended operation of the unit. The discovery was made by several Boeing engineers during a careful review of data generated by a National Transportation Safety Board test for the effects of thermal shocks on the PCU. . . . "This is the nature of our business," said Charlie Higgins, Boeing vice president/airplane safety and performance. "We identify very unlikely possibilities and take steps to eliminate them, or at least to further reduce their likelihood of ever happening. That's one of the ways we keep enhancing the safety of the aviation system."

It was too late for Cox to fly back to St. Petersburg, so he ended up at a Motel 6 near the Pittsburgh airport, watching reports on the rudder discovery on CNN. "This could wrap up 427 quickly," he said in a phone conversation between reports. "This could be the 'Aha!'" He said the lack of markings inside the valve was evidence that the gremlin in the rudder system "is a thief in the night. It leaves no trail."

He praised Boeing for being so forthcoming about the discovery. "I'm extremely pleased. They stepped forward."

Just before Thanksgiving, Phillips went back to the Parker plant in Irvine to compare the valve from the USAir plane with the ones from the Eastwind and the factory PCUs. He wanted to find out if there was something that made the 427 valve jam when the others would not.

Every rudder valve was slightly different. All valves had to meet certain Boeing and FAA standards, but the holes for hydraulic fluid on each one were never exactly the same. The tests so far suggested that some valves could be more prone to reverse than others.

At Parker, the three valves were each disassembled and examined and then hooked into a test rig to see how far off neutral they had to be moved before the rudder would reverse. The factory valve performed the best. It would not reverse until the outer slide was 38 percent extended. But the USAir and Eastwind valves would reverse more easily, when the slides were 12 and 17 percent extended, respectively. Also, a measurement of the distance between the valve slides found that the USAir unit was considerably tighter than the other two.

That was the final piece of evidence that Haueter had waited for. At last he had proved that the USAir valve was unique. After three years and hundreds of tests, he now had a scenario for what had happened to Flight 427.

It went like this:

It was a smooth flight from Chicago to Pittsburgh, so there was not much movement by the yaw damper. That lack of movement might have allowed particles to build up in the hydraulic fluid. There could have been a modest thermal shock to the PCU because of a problem with a hydraulic pump—not enough to set off a warning to the pilots but enough to send hot fluid rushing into the cold valve, making it suddenly expand.

The PCUs on other 737s might have tolerated that without trouble. But the valve on this particular USAir plane was especially tight. The thermal shock and the contaminants caused a jam. And the jam happened when one slide inside the valve was slightly off center and more likely to reverse. The pilot or the yaw damper was commanding the rudder to go right, but it went hardover to the left.

All of this occurred at the most vulnerable speed for a 737, when the plane was flying at 190 knots—the crossover point when a rudder hardover could not be countered by turning the wheel. The pilots compounded the problem when they pulled back on the control column, which made the plane lose speed and stall. The plane spiraled down and crashed into the hill in Hopewell.

It always takes a chain of events to cause a crash. In this case, it took the wake turbulence, the startling of the pilots, the fact that the plane was flying at the crossover point, the uniqueness of the valve, the jam and reversal, and the mistake of pulling back on the stick. If any one of those things had been different, the crash would not have occurred.

What the hell is this? Haueter thought he finally knew the answer.

He had moles in Boeing who gave him inside information about the company's strategy, alerting him when Boeing was preparing a full-court press to lobby the NTSB or when the company was softening its approach. The moles disagreed with the Boeing position of blaming the pilots, which Haueter found reassuring. He was glad that some Boeing employees had made their own decisions and believed the rudder had reversed. Haueter suspected there also were NTSB moles who told Boeing what he and Phillips were thinking. So he wasn't surprised when Howes, the Boeing coordinator for the investigation, called shortly before Christmas 1996.

"We understand there are people at the NTSB who, if they wrote the report today, would find fault with the airplane," Howes said. "We don't know how you can possibly say that. We'd like to know who these people are so we can help straighten them out."

"Rick," Haueter said, "I don't know how the report is going to go until it goes to the five board members. But many people have quite frankly told me that they think the airplane rudder system caused the accident—and several of those work for Boeing."

The Boeing team realized it was facing long odds. Haueter was still holding weekly conference calls with the parties, but they had less to discuss each week. After one call in early 1997, McGrew, Howes, and aerodynamic expert Jim Kerrigan sat in the Boeing air safety conference room looking glum.

The "Boeing Contribution" had been a flop. M-Cab had not been convincing. Haueter and Phillips wanted to blame the plane. It looked like Boeing was about to lose.

"It's a feeling of banging our heads against the wall," sighed Kerrigan.

They could not understand why Haueter and Phillips had become so narrowly focused on the plane. There were no conclusive data that showed there had been a jam—no scrapes, no marks. Yes, they had discovered a new failure in the rudder system, but there was no proof it had occurred on 427. Haueter and Phillips had a purely circumstantial case and had not seriously considered the circumstances on the other side, which suggested the pilots had caused the crash.

"You would hope the management of the NTSB doesn't form opinions ahead of time," Howes said, but everyone in the room knew that was wishful thinking. All three engineers thought the NTSB was obsessed with finding a cause for the crash and making sure the board did not end up with another unsolved case.

McGrew acknowledged that Boeing had made a few strategic mistakes in the investigation. The whole episode had been an education for him about the politics of aviation safety. He had always believed that all you needed to prove a point was solid data, but this investigation showed that the NTSB was getting more and more political under Hall and that Boeing had to play the political game if it wanted to succeed. McGrew said he had not done a good job of communicating with Hall and other NTSB officials along the way.

The realization that Boeing was about to lose was especially painful because so many employees had invested so much time in the investigation over the past two years. "People have burned a lot of hours on this," said Howes. "To have it come down to something we don't think is right is very discouraging."

Deep down, McGrew was convinced that the pilots caused the crash, but he didn't think there was enough evidence to list that as the probable cause. "We still cannot say we know what made the rudder go in. We think we know, but we'll never know for sure."

A friend called Brett the day after Boeing's announcement about the thermal shock discovery. "Hey, did you hear? They found the cause of Joan's crash."

Brett rushed out and bought the November 2 *Chicago Tribune,* but the paper did not have anything about the discovery. He found an item in the *Wall Street Journal* two days later, but it was a short story that did not capture the magnitude of the finding. He was losing hope that they would ever be able to prove what had happened. It was clear the rudder was to blame. Why couldn't the NTSB finally pin it on Boeing and USAir?

"I feel like Nicole Brown Simpson's relatives," Brett said. "There's this mass of information, but they can't get a conviction."

He had gotten a mixed response from entertainment companies when he made his pitch about T. Rex's Dino World Cafe. Everyone seemed to like the concept, but nobody would commit any money for a prototype. After several rejections from big companies, Brett decided to raise the money himself. He would give it a year. If he didn't succeed, he'd find a new job.

The possibility of getting rich from the restaurant was one of the driving forces for him, but he did not think of his lawsuit that way. It was a tool of vengeance. "The money doesn't mean that much. I got a fair amount from life insurance and that hasn't brought me any happiness."

He said he had finally realized that Boeing and USAir were both so huge that even a multimillion-dollar payment would barely cause a ripple in their finances. "There isn't enough money in the world to make them suffer," he said. In his view, Boeing and USAir were ducking responsibility in their court filings and public statements about the crash. Brett told Demetrio that in order to settle the suit he wanted an admission of guilt from the companies.

"You and I will be a fossil on another planet before that happens," Demetrio said.

On January 15, 1997, Brett happened to flip on the TV as he sat down to eat his lunch. A network was covering Vice President Al Gore's speech to a conference on aviation safety and security.

"As you know," Gore said, "the investigations into the crashes of Boeing 737s in Colorado Springs and Pittsburgh have not yet been closed. But those investigations have identified improvements that could help eliminate the chance of rudders playing a role in future accidents. These changes can and should be made without delay."

Brett listened intently. It sounded like the 737 was getting fixed.

"Boeing has developed modifications to the rudders of older 737s that will improve safety," Gore continued, "and they are going to begin retrofitting those planes, largely at their own expense, without waiting for a government mandate. Under a schedule to be developed by the FAA, these improvements will be made in the next two years. This is a major action: it affects some 2,800 planes worldwide, 1,100 of them here in the United States."

Tears welled up in Brett's eyes. It sounded as if the government was actually doing something to prevent another crash. He felt vindicated. He had believed all along that there was some kind of flaw in the rudder system. He felt as though he had reached the end of a long road.

At a news conference later that day, Boeing officials announced the specifics. The company would modify every 737 rudder valve so it could not reverse. A limiter would be installed on the PCU to prevent the rudder from going hardover. And, in a reminder that Boeing still believed the pilots were at fault, the company said it would pay for sensors on the rudder pedals. The next time a pilot stomped on the pedal by mistake, it would show up on the flight data recorder.

Haueter was disappointed that Gore praised Boeing repeatedly and made only a passing mention of the NTSB. Gore made it sound as if Boeing, out of the kindness of its heart, had generously offered to spend $150 million to fix the pcus. Never mind that the changes were a direct result of the work by him and Phillips!

"We dragged these people kicking and screaming for the last two years, and all of a sudden they are getting all the credit," Haueter said. "We have just been pounding them and pounding them, and now we don't exist."

His boss, Bernard Loeb, shared his anger. "Boeing didn't do this because their hearts told them to do it. Their lawyers told them to do it," Loeb said. "They didn't have a goddamn choice."

The same thing had happened two months earlier when Boeing made the dramatic discovery about the reversal. Whom did Boeing call then? The FAA. The company acted like the NTSB didn't matter. Haueter had called Howes, the Boeing coordinator, and hollered at him about the fact that Boeing had not notified the NTSB about the thermal shock finding. Howes had sat through the entire Halloween meeting without saying a word. But he insisted to Haueter that no one at Boeing had told him about it.

Haueter had discovered the downside of being the watchdog. To the public and the news media, NTSB investigators were crusaders for safety who could do no wrong. They were the guys in the white hats who found everybody else's mistakes. But when given the chance, the groups that the NTSB had attacked welcomed the opportunity to fire a few zingers back. So when Boeing and the FAA announced a safety fix, they rarely gave the NTSB credit.

The Boeing announcement was shrewdly timed. Haueter had been working on a new set of recommendations that called for immediate pilot training about the risk of a hardover. He felt that Boeing was trying to preempt him by announcing the rudder system changes first.

Boeing also managed to steal the thunder from *Dateline,* the NBC newsmagazine that had been working on a segment about 737 rudders for several months. A *Dateline* crew had been out to Renton and interviewed several Boeing officials. Boeing made sure that *Dateline* correspondent Chris Hanson got a chance to ride in M-Cab. Two weeks before Gore's announcement, rumors circulated through the investigators that *Dateline* was about to slam Boeing for the rudder problems. The segment was scheduled to air January 19.

But then came Gore's announcement on January 15, with Boeing saying it would improve the plane. That took the wind out of the *Dateline* segment, which ended up being a surprisingly positive piece that said the airplane was getting fixed.

GRUNTS

Malcolm Brenner, the eccentric NTSB psychologist, had made his name at the safety board by studying a drunken sailor. He and Jim Cash, the sound expert, had done a groundbreaking study of radio tapes of the Exxon *Valdez* accident that indicated the ship's captain was drunk. To show he was intoxicated just before the ship ran aground, they counted the number of seconds it took him to say each word. He slurred the phrase "Exxon Valdez," saying it 24 percent slower right before the accident than he had thirty-three hours earlier.

Speech analysis was still a very new tool for plane crash investigations. For a 1985 Japan Air Lines crash, Japanese investigators measured the pitch of the pilots' words to calculate when they were affected by stress after a sudden decompression. According to the investigators' calculations, a calm Japanese male spoke at a frequency of 150 hertz, but they found that the pitch of the captain's voice got as high as 410 hertz after the pilots heard a bang. The pitch of their voices also indicated the pilots were feeling the effects of hypoxia because they had neglected to don their oxygen masks.

Brenner decided to use the same technique on Flight 427's tape. The result might settle the debate over the Boeing theory that one of the pilots panicked because of the wake turbulence and slammed his foot on the rudder pedal. Brenner also hoped to look for correlations between the pilots' grunts and when the rudder moved, which might indicate when they had pushed on the pedals.

Brenner and three outside experts studied the cockpit tape one word at a time. Using Cash's WAVES program, they counted syllables and measured the pitch and volume of each sentence. Germano had said "four-twenty-seven" seventeen times during the tape, which gave them a consistent phrase to measure. As stress increases, humans raise the pitch of their voices. Anything above 300 hertz is considered screaming, an indication that the person has panicked.

They analyzed the way Germano said it each time. His average pitch was 144 hertz before the plane hit the wake turbulence, but it jumped to 214 hertz the last time he said it, which was about seventeen seconds after the wake turbulence. When they looked at his other comments on a bar graph, they saw a gradual increase in his frequency, which indicated that his stress level had progressively increased until he screamed and panicked at the end. There was no evidence that Germano had become "overaroused" when they hit the wake, as Boeing had suggested. They also found no evidence in Germano's voice that he was trying to forcibly move the controls during the emergency. That was an important detail because it focused attention on Emmett, who had been the flying pilot on the Chicago-Pittsburgh leg of the trip.

The results were less conclusive about Emmett. Scott Meyer, a navy sound expert, said Emmett's grunts indicated that he was straining, but it was not clear whether the straining came from pushing his feet on the rudder pedals or moving the wheel with his arms. The sounds of straining stopped once the autopilot was switched off, which allowed Emmett to turn the wheel more easily.

Alfred S. Belan, a sound expert from Russia who had analyzed hundreds of cockpit tapes, used a novel approach to figure out what Emmett did. He analyzed the word "shit."

Emmett had said the word during a calm moment about twenty minutes before the crash. He was trying to program the plane's navigational computer, but kept having trouble. "Aw c'mon, you piece of shit!" he said. He said it two more times—once as the plane was starting to point nose down and a final time just before the crash. Belan compared his breathing each time and studied color graphs from Cash's WAVES program. The graphs turned "shit" into orange lines that looked like a salmon filet, but Belan could look at them and see the difference between a grunt, Emmett's inhaling, and the word.

When people make a great physical effort, they take forced and rapid breaths. When Emmett said, "Aw c'mon, you piece of shit" to the flight-management computer, his breathing was normal. But after the plane was jostled by the wake, he grunted, and his voice showed signs of straining. When Emmett said the word seven seconds later, he said it softly and was not straining. That suggested he was no longer fighting the wheel or rudder pedals. It was easier to turn the wheel at that point because the autopilot had just been switched off. The final time Emmett said "Shit!" he was straining again, probably because he was pulling back on the control column, trying to pull the plane's nose up as the ground loomed closer.

Brenner thought that the sound analysis disproved Boeing's allegations that the pilots panicked early. The change in pitch was more gradual, proving that they didn't panic until well after the wake turbulence. Germano said "427 emergency!" to air traffic controllers seven seconds after the stickshaker went off, which Brenner regarded as a rational, constructive action—to let controllers know the plane was in trouble.

Brenner's biggest finding involved the grunts.

He put Emmett's heavy breathing and grunts on a time line and noticed that they perfectly matched the theory of a rudder reversal. They occurred during the crucial three-and-a-half-second period when investigators believed the reversal occurred.

Emmett's soft grunt at the beginning came four-tenths of a second after the rudder began to reverse, when he would have felt the pedal pushing back against his right foot. He grunted louder about one second later, the time when the rudder pedals would have fully snapped back against his pushing. It was a nearly perfect match that indicated Emmett had been struggling against the rudder pedal but couldn't stop it from reversing.

It was the best evidence yet that the rudder had reversed.

In January of 1997 Haueter had a difficult time sleeping. Many nights he found himself lying awake in the big four-poster bed that he shared with his wife, worried that another 737 would crash. The FAA had issued a slew of airworthiness directives that mandated changes to the 737 and how it was flown, including many that had been requested by the NTSB, but the agency had not acted on the NTSB recommendation that Haueter considered the most important—fixing the rudder's power control unit.

It had been two months since Vice President Gore had said Boeing would fix it, but Haueter had seen little action. Where was the urgency? Boeing and the FAA were taking a leisurely pace. The FAA—particularly McSweeny—often sounded like an apologist for Boeing. The FAA had mandated a weekly test for rudder jams, but the test was virtually meaningless. It would tell pilots if the PCU was jammed at that instant, but it didn't tell whether a PCU would jam in the future. Haueter thought that pilots needed to be warned more clearly about the problems.

It was time to light a fire under the FAA.

If Haueter had been a politician, he would just have held a news conference and spoken his mind. But that was not how the NTSB operated. There were unwritten rules of engagement about how to blast the FAA. It was done discreetly, usually by letter to the FAA administrator, which was conveniently faxed to aviation reporters around the country.

Haueter believed the 737 no longer met federal safety standards. He talked with colleagues about whether he should recommend that the entire 737 fleet be grounded but decided he didn't have enough evidence for such a drastic step. The plane's fatal accident rate was low compared with those of other planes—even if the rudder PCU did not meet federal standards. The odds of a jam and a reversal were still remote. Nevertheless, he was worried there might be another crash.

In the letter, he used strong words. Haueter had grown more self-assured in the two and a half years since the crash. He'd been promoted to chief of major investigations, which meant he was now the head of all the big accidents, including the ValuJet crash in Miami and the granddaddy of them all, TWA 800. He was not shy about blasting the FAA and Boeing when he felt he had to. The ten-page letter, signed by Hall and endorsed by the other four board members, laid out Haueter's case about how the soda can valve on the USAir plane was unique and more prone to reverse. The letter called on the FAA to alert pilots about the potential for a reversal and to speed up the process to fix all PCUs in the fleet. His words were surprisingly frank. He said the 737 was not as safe as other planes because the PCU was vulnerable to a reversal.

To make sure that the message got prominent play, Haueter, Loeb, and public affairs director Peter Goelz held a background briefing with reporters from the *New York Times,* the *Washington Post,* and the TV networks. Within hours, the news was spread to millions of viewers.

"Federal air safety experts are now satisfied they know what caused the mystery crash of a USAir Boeing 737," Dan Rather said a few hours later on the *CBS Evening News.* "It was the rudder." Without naming anyone at the safety board, CBS reporter Bob Orr said investigators "believe a malfunction in the Boeing 737's rudder system rolled the plane into a fatal dive. And they suspect a similar failure caused the crash of a United Airlines 737 in Colorado Springs in 1991."

NBC anchor Tom Brokaw said, "There may be an answer tonight to questions that for years have perplexed investigators of two deadly plane crashes. That answer is certain to raise new fears about the world's most widely used airliner, the Boeing 737."

Haueter watched the NBC report at home and felt tremendous relief. His worries about the plane had been told to the world. He wasn't holding a secret anymore. Everyone knew what he knew.

In the meantime, Boeing had gone into damage control mode. It faxed a statement to the press that said: "Boeing is and has been working with suppliers on an already aggressive schedule" to fix the rudder system. The changes "will serve to make a safe airplane even safer."

That had become the 737's mantra, chanted by people at Boeing, at the airlines, even at the FAA. It was the perfect phrase to put a positive spin on the message. The 737 wasn't dangerous. They were simply making it better.

Despite heavy news coverage about the NTSB's letter, there was barely a ripple of reaction from the public. The *New York Times* reported "a collective shrug of indifference" from passengers. "Travel agents, corporate travel managers and other industry officials said yesterday that most passengers demonstrated unshaken confidence in the overall safety of air travel," the *Times* said, "and appear to feel that the Government would have grounded the nation's fleet of 1,100 737's if they were truly dangerous."

Boeing's deft response minimized the repercussions for the company. Its stock price had not been affected by any of the 737 announcements. Wall Street didn't seem to notice what the NTSB said. Airlines seemed satisfied that Boeing was fixing any problems that the plane had, and they continued to flood the company with orders for new 737s.

Boeing wanted more tests that would push a 737 to its limit. A Boeing test pilot would fly over the Pacific Ocean and try rudder hardovers and other maneuvers to gather new data about the crossover point. Cox thought the tests were unnecessary and were being done to help Boeing defend itself from lawsuits. Worse, Cox was concerned that there might be an accident, which would lead to the immediate grounding of the entire 737 fleet. He joked that 737s would make good restaurants, which was a good thing because that's all they would be useful for if a Boeing test pilot dropped one into the Pacific.

But when investigators arrived in Seattle for the tests in the first week of June 1997, Boeing shocked everyone. For the first time, company engineers acknowledged that a jam and a reversal in the PCU could match the kinematic estimates for Flight 427. Until that point, Boeing had adamantly resisted any suggestion that a reversal caused the crash. The Boeing position had been firm: The pilots screwed up. But new estimates by Boeing's engineers provided a surprisingly close match between Flight 427's rudder and results from the jam/reversal tests. Boeing stopped far short of saying the plane was to blame, but just an acknowledgment that it was *possible* was a huge step. It was like Ronald Reagan saying nice things about communism.

Boeing's new position won Haueter's admiration. "Holy mackerel, I'm shocked," he said. "They have been fairly up front lately."

Cox also praised his rivals from Boeing for being so honest with data that appeared to indict their product. "The entire week was a night-and-day difference," Cox said. "They were very forthright, there was not any of the partisanship."

The flight test was uneventful. Boeing pilots compiled new data about the crossover point and hardovers and did not crash in the Pacific. But the most revealing event came when the plane returned to Boeing Field for a ground demonstration of a rudder reversal. The plane, which had just come off the assembly line for Southwest Airlines, was fitted with the special rig in its tail to jam the valve.

Haueter had asked for the demonstration to show what a rudder hardover would have felt like to Emmett and Germano. It took two steps to cause the reversal. You pressed down with one foot to move the valve's secondary slide into position, and then you had to stomp quickly with the other foot, which resulted in the second rudder pedal's snapping back, as if it was trying to throw your foot off.

It was a warm, sunny day in Seattle, once again proving that the city's reputation for rain was a myth perpetuated by natives who didn't want anyone else to move to their beautiful city. It was so bright that people standing on the pavement beneath the plane had to wear sunglasses. In the cockpit,

Brenner sat down in the right seat beside Mike Carriker, the Boeing test pilot. The six-foot-three Brenner was the same height as Emmett, and he had to put the seat all the way back to be comfortable. He slowly pushed each rudder pedal as far as it would go, as if he were doing a routine check of the rudder before takeoff. He then jammed his foot on the left pedal, causing the reversal. The left pedal began coming up, pushing relentlessly against his foot, fighting him all the way until it reached the upper stop. He did the test a dozen more times and found that the rudder pedal overpowered him every time. He was amazed by the force. In a fight with a jammed PCU, a human could not win.

Brenner had spent two and a half years trying to understand what had happened in 427's cockpit. He knew the pilots so well that he knew their pant sizes. He knew their allergies, their marital histories, and what they ordered from room service. He had heard the cockpit tape hundreds of times, had spent hours in M-Cab and weeks poring over the flight data. But not until this moment, when the pedal came snapping back, did he truly understand what had happened. It was creepy. Brenner felt as though he was reliving those horrifying seconds at 6,000 feet. For a brief moment, he became Emmett. *Oh, yeah, I see zuh Jetstream.* He pushed on the rudder pedal to recover from the wake but felt it snap back. *Oh shit.* It was unrelenting. No matter how hard he pushed, the pedal *Ohhh shiiiiiiittttt* kept fighting back as the plane rolled to the left and then spun out of control *God!* toward the gravel road in Hopewell. *Nooooo.*

Brenner was convinced. The rudder had reversed.

Phillips had a similar reaction. He was still the most cautious investigator at the safety board. Long after Haueter and others had become convinced that a reversal in the PCU had caused the crash, Phillips was reluctant to draw a conclusion. Some people at the board thought he was a little too cautious and too protective of the airlines and the manufacturers. But after trying the demonstration and seeing Boeing's new estimates, even Phillips seemed convinced that there had been a reversal.

Once again Haueter was working late.

Since his promotion, he'd gotten a better office with a sign beside his door that said NO WHINERS, but he had not bothered to get a nameplate with his new title. He had written his name on a piece of paper and taped it over the name of his predecessor.

He still had responsibility for the USAir investigation and often worked into the night. His wife, Trisha Dedik, had grown accustomed to his calls saying he would be an hour or two late, but she could never be sure exactly when he would get home. John Purvis, the head of accident investigations for Boeing, had just called their house to look for him. Dedik told Purvis her husband was still at the office, but then Purvis had called again, still looking for him. Dedik was fed up with the Boeing guys. They called at all hours and made it sound like everything was urgent. They treated her like she was Haueter's secretary.

Though he had called Dedik to say he was headed home, Haueter ended up working for another hour before calling again to say that this time he was *really* leaving. Dedik was furious. She took a book into bed and read until she heard the thunder of their garage door when he arrived home. She switched off the light and pretended to be asleep. He came into the room and kissed her on the shoulder, but she was too mad to face him.

The next morning, she told him to call Purvis in Seattle, where it was three hours earlier.

"Trisha, it is five-fifteen or five-thirty in the morning out there," Haueter said.

"So? He doesn't care. He is calling here at ten-thirty or eleven o'clock at night. Call him right now."

"Trisha, I'm not going to call him."

Some days she wished that he would leave the NTSB and find a job that was less demanding, where he didn't have to carry a beeper and work so late. She had never understood the whole preoccupation with beepers. Her office had given her one, but she stuck it in a kitchen drawer and promptly forgot the number.

"If I had to reach you, how would I?" Haueter asked her. "I can't beep you. You can always beep me."

"If it's really important," she said, "someone will find me."

As the fourth anniversary of the crash passed, Haueter began plotting his strategy to convince the board members to approve a final report that would say the rudder had reversed. He knew that Bob Francis would be his biggest hurdle.

Francis, a tall, balding man known for wearing identical blue oxford shirts and khaki pants every day, was a former FAA official who had become something of a national celebrity after his nightly briefings on the TWA and ValuJet crashes. He considered himself the diplomat of the safety board and said he liked to work out disagreements behind the scenes. But the investigators disliked him because he did not attend their briefings or read their reports before he spoke to the press. They believed he was weak with the FBI during the TWA probe and had become too close to Boeing.

Francis had disagreed with Haueter about several of the 737 safety recommendations over the years and had not signed off on them until Haueter agreed to soften the language. In this case, Francis was not convinced that Haueter had enough proof to blame the 737 for the crash.

"I think he'll fight us," Haueter said. "For some reason, Bob does not like to be controversial with industry. He just backs away. He says he is concerned with the credibility of the agency and wants to make sure that we work with people and try to get them to come along. He thinks we should do more things with gentlemen's agreements. Unfortunately, gentlemen's agreements don't work."

Still, Francis had a history of ultimately voting with the other board members, even when he disagreed. "While on one hand he seems to favor

Boeing over us," Haueter said, "in the final analysis, he has always signed the bottom line to go with the staff."

John Goglia, the most colorful character on the board, had decided not to vote in the Flight 427 case. A burly Bostonian, Goglia was a former USAir mechanic who had represented his union in the investigation until he was appointed to the safety board by President Clinton. He said he wanted to avoid any appearance of impropriety. "I'm controversial enough as it is," he said.

Haueter was worried that the board might reject his report or go with a weak probable cause. If that happened, he had a secret weapon: the families.

The 1996 ValuJet crash had shown how powerful they could be with the safety board. At the final meeting on the crash, the board was minutes away from adopting a probable cause that blamed the FAA and a maintenance company when Goglia suddenly recommended blaming the airline too. The family members—many holding big photographs of the crash victims—applauded and cheered. The emotional demonstration pressured the board to adopt Goglia's recommendation. Haueter figured he could use the USAir families the same way—if necessary.

If the board still balked, he planned to quit. He'd gotten three job offers from engineering and aviation companies recently and had turned them all down. He and Trisha didn't want to move away from Washington, and he did not want to desert the NTSB during its toughest investigation. But the offers had shown that he was valuable, and he was willing to take a new job if the board snubbed him.

23

DELIBERATIONS

Boeing and ALPA got one last chance to take shots at each other in their final submissions to the NTSB. The submissions were like closing arguments in a criminal trial. Once they were turned in, Haueter and his bosses would largely cut off communications with the parties. The report would be written in private, and the parties would probably not see it until the final board meeting. The investigators were like cardinals meeting in the Vatican to pick a pope. No one would know the outcome until white smoke poured from the NTSB chimney.

In previous crashes, the submissions had been long, boring letters written in the language of engineers. But Flight 427 was different. The parties were targeting not the NTSB investigators, who had pretty much decided to blame the airplane, but the five board members who would vote on the case. The submissions also had a broader audience: the public and the courts. Each side wanted to reassure customers and juries of its innocence.

Haueter viewed the submissions as an important part of the contentious party system. He knew they would be filled with bias, but he felt they brought a healthy discussion and helped the NTSB sort out the facts. He was eager to see them because he had heard there were internal fights at Boeing and at ALPA.

His moles at Boeing said there were two distinct camps within the company. One wanted to lay the blame squarely on the pilots, saying there was no evidence that the plane had malfunctioned. The other camp was more cautious and wanted to admit that there was no conclusive proof either way.

He also heard of a similar fight within ALPA about whether to mention the mistake the USAir pilots had made in pulling back on the stick. One group wanted to ignore that fact to avoid conceding that the pilots made a big error. The other camp wanted to acknowledge it and then show that the pilots were doing their best, given that they had no training about rudder problems or the crossover point.

The sixty-eight-page ALPA submission had no charts or drawings, but it presented a clear, point-by-point case to show why the plane was at fault. It portrayed Emmett and Germano as helpless victims of a sudden malfunction. It said they fought to regain control but could not overcome the problems of the 737. The union's report made a circumstantial case that the valve had jammed and reversed. It conceded that there were no marks to prove a jam but said the thermal shock test showed that some jams did not leave a mark.

"The situation was perilous," ALPA wrote. "The more the aircraft turned to the left, the stronger the first officer's tendency to apply increased right rudder pedal pressure; the harder he pushed on the right rudder pedal, the more certain it became that the jam would not clear."

In building its case, ALPA agreed with Brenner's interpretation of the first officer's grunts, saying there was a distinct correlation between the grunts and when the rudder would have reversed. But ALPA stopped short of saying the plane should be grounded. Cox and other ALPA pilots continued to say that a rudder hardover was "a low-probability event." The union also dodged the touchy question of the pilots' pulling back on the stick. It was mentioned only briefly in the report and was not characterized as a mistake.

The submission from USAir also blamed the plane. Relations between the airline and Boeing had been strained, especially since USAir had decided to buy a fleet of new planes from Airbus, Boeing's archrival. In its submission, USAir complained that Boeing had not warned anyone about the risks of the crossover point: "Under any circumstances then known to the airline industry, the actions of the crew of USAir Flight 427 were reasonable and correct. . . . The crew had seven seconds, at most, in which to recognize, analyze and recover from a previously-unknown malfunction."

The Boeing submission came in a glossy spiral binder and was professionally typeset with headlines and subheads. There were foldout charts of the flight data, drawings of the rudder system, and a table that compared the evidence for pilot error against the possibility of a PCU jam and reversal.

The eighty-three-page report appeared to be a compromise of the two camps within Boeing. It was not as absolute and righteous as the company had been a year earlier with the "Boeing Contribution," the blatant lobbying attempt that had been such a flop. Whereas that document had been unwavering in its conviction that the pilots were to blame, the new report conceded that there was no proof: "In Boeing's view, under the standards developed by the NTSB, there is insufficient evidence to reach a conclusion as to the cause of the rudder deflection."

One section cautioned the NTSB about trying to name a probable cause when there was insufficient evidence. It said there had been a "clamor for a definite and expeditious explanation" and urged the safety board to be careful. "In order to avoid the wrong answer, it is essential that any cause identified by the Board in this accident investigation be supported by facts and evidence. Mere suspicion, inference and conjecture must not suffice." It quoted Hall as saying that the only thing worse than not being able to solve a case would be to come up with the wrong solution.

The Boeing submission said there was no connection between 427, the Colorado Springs crash, and the harrowing incident on the Eastwind plane. According to Boeing, a new computer analysis of Colorado Springs confirmed what the company had maintained all along: A powerful wind had thrown the plane to the ground. Boeing dismissed Eastwind as a minor event caused by a rudder device that was misrigged and said there was no proof that the Eastwind valve had jammed, just as there was no proof in the USAir valve.

The Boeing report seemed to have been written largely by the engineers who wanted to blame the pilots. It said the NTSB did not have enough evidence to assign a probable cause—unless the safety board wanted to blame the pilots. In that case, there was plenty of evidence.

The report offered a strong defense of the 737 rudder system and explained why it would not have malfunctioned on Flight 427. Nearly every place that the report mentioned the possibility of a jam, it used the word "hypothetical." ("Hypothetical scenarios exist that would provide a full rudder deflection to blowdown. However, very specific conditions are required for each hypothetical failure scenario.") The word appeared at least nineteen times.

But when Boeing discussed pilot error, the word "hypothetical" was not used. The report said airline pilots were often startled by wake turbulence, overreacted, and stomped on the rudder. That was a fact. In Boeing's view, pilot mistakes were not hypothetical.

To show how a pilot could keep his foot on the pedal, Boeing cited the problem of "unintended acceleration," when a driver of an automobile mistakenly stomps on the gas pedal instead of the brake, a mistake that has destroyed front windows in many 7-Elevens.

The company offered a scenario showing how the pilots could have caused the crash: Emmett was so relaxed before the wake turbulence that he referred to the Jetstream in "a drawn-out, feigned French accent." When the plane was jostled by the wake, Emmett pushed on the rudder pedal once or twice to level the wings. He was so startled that he did not realize that he kept pressing on the rudder. He then pulled back on the stick, which stalled the airplane and made it crash.

Boeing did not discuss the dangers of the crossover point, or the fact that the pilots had not been warned about it. But the company's submission said that Emmett and Germano could have recovered by simply turning the wheel to the right.

The message was subtle but clear: The 132 people on the plane would be alive today if the pilots had done the right thing.

In a surprise to everyone in the investigation, the FAA chimed in with a submission a week after Boeing did. The FAA was always a party in NTSB investigations, but it rarely made a submission. Its relations with the safety board were rocky, so FAA officials were selective about when to pick fights. In most investigations, the FAA sat quietly and took its licking when criticized by the NTSB. But this time the agency adopted an unusual tactic: It told the NTSB there wasn't enough evidence to name the probable cause.

"The FAA, upon review of the evidence, cannot conclude that a failure mode . . . has been identified. Any causal findings, to be legitimate, must have conclusive evidence to support findings of a hardover or reversed rudder. Such evidence has yet to be found."

Haueter was struck by how much the FAA submission sounded like Boeing's. In his view, the FAA was just posturing to protect itself. If the NTSB blamed the airplane and said it didn't meet federal safety standards, everyone would want to know why the FAA hadn't grounded the plane. He thought the FAA was trying to preempt the NTSB by saying there was insufficient evidence.

Haueter believed he had enough evidence. No, he didn't have absolute proof that the valve had jammed. But he had an extremely strong case that it was the most likely cause. Now he had to convince the most important audience: the board members.

The report was marked "DRAFT—Confidential" when it went to board members in early February 1999. It was five hundred pages long—so big that it had to be held together with a rubber band instead of the usual binder clip. It was the longest crash report in the NTSB's history.

The probable cause statement was succinct. It said the pilots lost control because the rudder reversed. The report was sharply critical of the 737, saying the plane was not as safe as it should be. Boeing was fixing the rudder valve to prevent a reversal, the report said, but that was "not an adequate fleet-wide remedy." The report warned that the plane was still vulnerable to rudder malfunctions that could have "catastrophic results," and it said there could be additional 737 rudder problems that had not been discovered.

The report said the FAA should replace the unique rudder valve with one that is "truly redundant," which could be done by adding a second valve or splitting the 737's single rudder panel into two.

The report was very much a group effort. Haueter and his investigators had written long sections that had been compiled and edited by an NTSB report writer. But the person with the greatest influence over it was Bernard Loeb, Haueter's boss.

Loeb was an intense, opinionated man who had become the NTSB's director of aviation safety midway through the investigation. He had an in-your-face style and wasn't shy about telling people they were wrong. Peter Goelz, the safety board's managing director, recalled that once, during an argument about Flight 427, Loeb had called him an idiot. Loeb was renowned for being a micromanager. George Black, an NTSB board member, jokingly called Loeb and his staff "the Borg," after the evil force on *Star Trek* that controls the brains of drones.

Some people at the NTSB thought that Haueter relinquished too much power to Loeb and didn't challenge him enough. Haueter himself was frustrated that Loeb had such a heavy hand, but Haueter felt that he went as far as he could. He stood up for the ideas that mattered, but ultimately he had to respect that Loeb was his boss.

Indeed, Loeb's aggressive style was valuable. Haueter's "Holy Mackerels" and his friendly disposition sometimes made him seem overly tentative. Loeb became the strong advocate the investigation needed at the end, someone with status and a loud voice who could get the final report approved. Loeb's style was in keeping with the safety board's culture of argument. The shouting matches got all the conflicts out into the open. A weak theory—or a weak investigator—didn't last long.

Loeb was pushing a theory that was like a three-legged stool. He thought that, individually, none of the three incidents gave the NTSB enough evidence to blame the problem on a rudder reversal but, collectively, the three provided enough proof that the rudder had jammed and reversed.

New simulations by NTSB computer whiz Dennis Crider showed that a reversal would explain all three incidents. Crider's most important finding was this: The rates of rudder movement on the three planes were nearly identical. That was powerful evidence. It was highly unlikely that three pilots on three different days in three different airplanes would move the rudder at exactly the same speed.

Loeb and Haueter now were focused on counting votes. Goglia had disqualified himself from voting, but they still needed three of the four remaining board members. If Francis voted against them, they needed everyone else.

Haueter was nervous that they wouldn't get the votes. He had heard rumors that the board members were split 2–2 on whether to blame the plane. They were said to be skeptical about the report and wanted to blame the crash on "undetermined reasons."

Undetermined reasons.

Those were the big blue words on the cover of the Colorado Springs report, and Haueter winced at the thought of them on *his* report. If they appeared on the Flight 427 report, Haueter feared he would be remembered "as the guy who flubbed it." Even worse, the 737 would continue flying with its elusive rudder problem, and the NTSB would have no leverage to force Boeing to fix it.

Haueter was concerned that Crider's findings were not having an impact with the board. Crider had to keep revising them as he made last-minute discoveries, but the numbers always added up to the same conclusion—a rudder reversal. Haueter was afraid it looked to the board members like the investigation was in chaos and was biased toward that conclusion.

Haueter and Loeb were also frustrated by Phillips, who was reluctant to blame the crashes on rudder reversals. Phillips said the reversals were a plausible explanation, but that he didn't have enough solid evidence to be definitive. "I can't lay a part in front of you and say this is what broke," he said. He felt the NTSB lost credibility if it pushed too hard for a cause when the evidence wasn't solid.

Haueter thought his friend was being wishy-washy. He said Phillips would go out to Boeing and come back sounding as though he had been brainwashed. He kept telling Phillips that they had a strong case and had some latitude because they only had to come up with a *probable* cause.

As Crider scrambled to finish the new computer simulations, Boeing, which knew all the angles to work at the NTSB, was busy lobbying the board members. Boeing had taken each of the board members for a ride in the M-Cab simulator to show how easily Flight 427's pilots could have saved the airplane. The company also had provided the board members with a video about the plane that discussed the extensive plans for safety improvements.

USAir was doing its own lobbying, trying to persuade the board members that the plane was at fault and thus putting the airline in the odd position of arguing that every 737—including the 200 that the airline itself operated—had a safety problem. A USAir official came to the NTSB offices and showed the board members a horrifying video that combined a computer animation of the crash with the actual cockpit voice tape. It gave board members a chance to see and hear two pilots fight and scream and then die, to emphasize USAir's position that the pilots had no idea what the plane was doing.

Haueter and Loeb figured they had a good chance with two of the board members—Chairman Jim Hall and John Hammerschmidt. Hall did not have the technical background to understand the intricacies of the valve, but he had been suspicious of Boeing and seemed to have high regard for the staff recommendations. Hammerschmidt, a shy man who was virtually invisible as a board member, was also expected to go with the staff recommendation.

Bob Francis was shaping up to be just as difficult as Haueter had feared. He said the report was too absolute about the rudder reversal and that it was overly critical of the FAA. In his view, the NTSB had "fairly shaky evidence" and should not be so critical. He met with Hall in the chairman's office and said he would vote against the report unless it was toned down.

The other key board member was George Black, a brainy highway engineer from suburban Atlanta. He spent far more time studying the evidence than the other board members and filled a spare office with engineering reports, maps of the plane's radar track, and a small plastic model of a 737 that he used to demonstrate the crash. He scrawled a sign for the door that said, THE WAR ROOM.

No one was sure how Black would vote. He liked to play devil's advocate, throwing out new ideas that often contradicted each other. He liked to tease Boeing executives about the possibly dismal future for the 737, their best-selling product, if the rudder was blamed for the crash. "Boeing has no corporate sense of humor," he grumbled.

Black was convinced that the plane had a problem in its rudder system. He had discussed the crash with pilots and engineers and decided that Emmett and Germano would not stomp on a rudder pedal and hold it to their death. He was especially impressed with Crider's simulations and Brenner's work matching the grunts to the precise moments at which the rudder appeared to reverse. He was not persuaded by Boeing's M-Cab ride and thought it was understandable that the pilots—unaware of the dangers of the crossover point—could lose control of the plane.

Yet he thought the report was too strong. "Tom and Bernie come in here thumping the report like it were the Torah," he said, but they still had no proof. He said the board members had some latitude because they were naming the *probable* cause, but he wondered, "At what point does this rise to the level 'probable'?" He also was concerned that Crider's simulations were a little too perfect, matching the reversal scenario every time, even when Crider had to make adjustments. Haueter assured Black that the reversal scenario was just one of several that could match, but Black was still wary. He worried that Loeb and his deputies were too adamant about the plane.

Suddenly, right in the midst of the deliberations, there was another 737 scare: a USAir MetroJet plane had a strange rudder incident.

Shortly before noon on February 23, 1999, MetroJet Flight 2710 was cruising at 33,000 feet over Maryland when the pilots noticed the wheel suddenly turn to the left. They quickly realized the autopilot was turning the wheel to compensate for rudder movement. But the pilots had not touched the rudder pedals. The copilot then put his feet on the pedals and discovered that they were displaced to the right. He turned off the autopilot and pushed on the left pedal to return the rudder to the center position, but the pedals seemed to be jammed.

The pilots quickly followed an emergency procedure that had been developed by Cox and another USAir pilot. They turned off the yaw damper, the device that makes small adjustments to the rudder, and activated the standby rudder system, which uses a backup hydraulic valve instead of the main one. The rudder pedals moved back to the center.

The pilots announced to passengers that they had a flight control problem and then landed at Baltimore-Washington International Airport. After they parked at the gate and everyone got off, a platoon of investigators from the safety board, the FAA, ALPA, and USAir arrived and began removing evidence. They took the two flight recorders, the yaw damper coupler, the rudder PCU and its valve-within-a-valve, and one liter of hydraulic fluid.

The flight recorder showed that the incident was very different from the Pittsburgh crash. Instead of the fast-moving hardover, the MetroJet rudder had moved slowly to its limit. One FAA official called it "a slow-over." The incident was even more curious because the plane had one of the new, improved rudder valves that prevented a reversal, and the valve showed no signs of a jam.

The incident left Haueter and Loeb scratching their heads. It added urgency to their work in the final weeks before the board meeting, but it also raised some troubling questions. Had they found all the problems that might lurk inside the rudder system? Did the valve have a different flaw that they had not uncovered? Had there been a jam somewhere else in the PCU? Cox, who had blamed the USAir crash on a "gremlin" in the plane, said the MetroJet incident showed there was "another gremlin in the tail of the 737."

It also revealed a major shortcoming of all their work. Haueter's team had some of the smartest engineers in aviation, supplemented by the Greatest Minds in Hydraulics. Boeing had deployed its best and brightest and called in people with decades of experience. But for all that brainpower, they still had not come up with a definitive answer about what had happened to Flight 427. That was remarkable because they were not dealing with a million lines of computer code or a newfangled electronic gadget. They were dealing with an old mechanical device—an ordinary valve designed when John F. Kennedy was president. And now, more than thirty years later, they were struggling to understand the gadget and the myriad ways it could work.

"I don't even think the inventors understand it," said Steve O'Neal, the FAA flight test engineer. He said the MetroJet incident suggested that more changes were needed in the rudder system. "We just hope another 737 doesn't come screaming out of the sky in the meantime."

George Black, the NTSB board member, was worried that the MetroJet incident revealed that the investigators had been too fixated on the valve and had missed a problem elsewhere in the plane. He knew they had ruled out hundreds of possibilities—everything from bird strikes to the fat guy theory—but he was still worried that they were missing the real problem. He was concerned that they would approve the report and then end up having to revise it later. "Are we premature?" he asked. "We want to make sure we don't start off down some path and decide it was unnecessary."

As they debated the wording of the report, Black and Francis were particularly concerned about one of its findings, which lambasted the FAA for certifying the 737 rudder system. The finding said that the 737 would not have been certified if the FAA had insisted on "a high level of safety."

If that was true, the board members said, the NTSB should be calling for the entire 737 fleet to be grounded—which would be viewed as a ridiculous request. The plane had 92 million flight hours and one of the lowest crash rates of any transport jet.

Black and Francis also complained that the report's recommendations for the rudder system were too specific, reading like a mandate for Boeing to split the system into two separate valves. They said the NTSB should not tell Boeing or the FAA how to design airplanes. The board should recommend broad principles and leave the details to the experts. Black kept invoking a philosophy used by physicians: "Do no harm."

Haueter and Loeb insisted that they did not want to ground the plane, which would put entire airlines out of business and wreak havoc with the world's economy. But they argued that the rudder system needed to be redesigned because a single failure could cripple the plane. Indeed, two years earlier, the five board members had unanimously approved a safety recommendation letter that said the 737 was not as safe as other planes.

While the other board members debated the wording, Hall began to strategize how he could get a unanimous vote. He believed a 4–0 decision was critical in order to maintain the board's credibility on such a touchy issue. Hall was comfortable with the strong wording in Loeb and Haueter's report, but he was perfectly willing to tone it down to get a 4–0 vote.

He was in a bind. If he kept Loeb's strong language but ended up with a 2–2 deadlock, the whole four-year investigation would go down the drain.

The FAA and Boeing might end up doing nothing to fix the plane. On the other hand, if the report was weakened too much, there was a risk that Loeb would get angry and publicly criticize the board. Loeb had never explicitly made such a threat, but he didn't have to. He was about to retire and had nothing to lose. He was well regarded among aviation reporters and a complaint from him that the board wasn't tough enough would surely make front-page headlines.

The deliberations showed the NTSB as a dysfunctional family. Francis rarely spoke with other board members or with the investigators, yet he complained that Loeb wouldn't permit his underlings to speak freely. But Loeb and the staff complained that Francis was aloof and had not spent much time studying the accident. Likewise, relations between Black and Loeb were strained. Black felt that Loeb was too abrasive and that he stifled discussion of ideas.

Hall appointed his assistant, Deb Smith, to act as a peace negotiator. She shuttled back and forth from Francis to Black to Loeb, trying to broker a compromise. Black and Francis had tremendous leverage in getting the changes they wanted. Francis, using his special counsel, Denise Daniels, as his own peace negotiator, sent a lengthy memo that detailed his concerns. He said he would vote against the entire report unless the changes were made. His complaint was primarily about tone. He wanted the report softened so it did not sound so absolute that the rudder had reversed.

There were lengthy debates about a single word. The draft report had said the 737 needed a rudder system that was "truly redundant." Loeb and Haueter had added "truly" to strengthen the sentence so McSweeny, the FAA safety official, would not have any wiggle room. They were afraid if they simply said the valve was not redundant, McSweeny would retort that it *was* redundant. By adding "truly," they gave the sentence more impact.

Black, who was often annoyed by Loeb's aggressive style, was not about to accept Loeb's word. Black suggested "reliably redundant," which ultimately was adopted as the final wording.

The group also debated whether to ask the FAA to establish an independent panel to assess the 737's rudder system. The panel, which would have representatives from Boeing, the FAA, the NTSB, and academia, would conduct a year-long examination and make recommendations on how the plane could be improved. Loeb regarded that as a waste of time. The NTSB was supposed to be *the* independent safety agency, he said. There was no need to call in another independent group to validate their work. That just undermined the NTSB's authority. But Black and Francis liked the idea of an independent panel that would take a broad look at the 737's problems and make an impartial recommendation. The suggestion for the panel got added to the report.

After exchanging drafts by E-mail, the board members also decided to tone down the probable cause statement. Instead of Loeb and Haueter's definitive assertion that a rudder reversal had caused the crash, the board members softened it to say the sudden movement by the rudder was "most likely" caused by a jam and reversal.

As the final meeting neared, it appeared that Smith and Daniels had brokered a compromise. But as they walked into the hotel ballroom, they were nervous that the deal might unravel.

The final meeting on Flight 427 was held at the Springfield, Virginia, Hilton on March 23 and 24, 1999. The big room looked like a movie set. Blue lights illuminated the curtains. The board members sat behind a wooden desk like a jury deciding if the 737 was guilty of murder.

Chairman Hall opened the meeting by reading a statement that said the event was part of the Government in the Sunshine Act, which required federal agencies to do their work "in open session." But in fact, all of the real debate had occurred behind closed doors long before the meeting.

It was one of the biggest days in Haueter's career, and his wife, Trisha Dedik, had come to watch. Haueter hadn't slept well the night before the meeting, waking up several times worrying about the outcome. But as he tossed in bed, he couldn't pin down any single thing that was likely to go wrong.

The ballroom was packed with about three hundred people. Relatives of Flight 427's passengers sat in a special section, many wearing photos of the victims on buttons or chains around their necks. Brett Van Bortel took a seat in the last row of the family section. He didn't realize it, but sitting in the row directly behind him were the people he blamed for Joan's death—top officials from Boeing and USAir.

Brett was looking healthy and confident. It had been four and a half years since the crash, and he had healed as much as anyone could. He was dating again and engrossed in his job at a mutual fund company. He listened attentively as Haueter began his presentation.

"Today, the investigative staff is pleased to present the report on the crash of USAir Flight 427," Haueter said. He noted that it had been the longest investigation in the safety board's history.

Over the next several hours, Haueter and his team explained the 737 rudder system and why they believed it had reversed on the USAir, United, and Eastwind planes. They spoke in absolutes, as if there was no doubt about what happened.

"The pilot is surprised and pushes harder—as hard as he can," Malcolm Brenner said about Flight 427. "But instead, the controls reverse and move the rudder all the way to the left."

Brenner talked about the risks of the crossover point. He said Emmett, Flight 427's first officer, discovered "for the first time in his career" that he could not turn the wheel to bring the wings level and stop the plane from rolling out of control. The veteran pilot was helpless as the big 737 began to plummet toward the ground.

Crider, a goateed former McDonnell Douglas engineer who was considered a genius with flight data, used so much jargon in his presentation that many people in the audience had no idea what he was saying. But they understood when he summarized the 737 incidents by saying, "A rudder reversal scenario will match all three events." In the audience, Boeing engineers could only listen and bite their lips. They had no opportunity for rebuttal. This was the NTSB's show.

Smith and Daniels sat behind the board members wondering if their compromise would fall apart. But as the board questioned Haueter and his staff, there were no hints of any disagreements. As the session stretched into late afternoon, Hall decided to adjourn and resume the following day, although many NTSB staff members wanted to wrap it up that evening.

The next morning, Haueter was feeling confident as he drove his Toyota 4-Runner to the hotel. "Everything we wanted to say is now out there," he said. "I feel pretty good about that." But as he pulled into the hotel parking lot, Haueter could see that his message about the 737 was not getting the attention he had hoped it would. There were only two TV trucks in the lot. Many network crews had been diverted to the Pentagon because U.S. planes had begun bombing Yugoslavia a day earlier. Hall's decision to extend the meeting to a second day also diluted the impact, because reporters had to write two incremental stories instead of a single, more powerful account. In many newspapers the 737 decision would be relegated to a short story buried inside.

When Hall reconvened the meeting, board members asked the investigators a few more questions. The questions indicated that they had no major disagreements with the report.

Hall then read the conclusions, the thirty-four findings that built the case for the probable cause. The entire room—the Boeing engineers, the USAir officials, the reporters, Brett and the other family members, Haueter and his staff—had waited nearly five years for this moment.

> The USAir Flight 427 flight crew was properly certificated and qualified. . . . No evidence indicated any preexisting medical or behavioral conditions that might have adversely affected the flight crew's performance.

The first set of conclusions dealt with what *had not* happened, to show that the NTSB had ruled out bombs, birds, and a midair collision. The conclusions were designed to satisfy the conspiracy theorists who still believed someone had blown up the plane.

> USAir flight 427 did not experience an in-flight fire, bomb, explosion or structural failure.

Hall moved on to the findings that built the case for the valve reversal. He read Conclusion No. 8, which even Boeing could support:

> About 1903:00, USAir flight 427's rudder deflected rapidly to the left and reached its left aerodynamic blowdown limit shortly thereafter.

The next finding ruled out pilot error. All of Boeing's lobbying—the hours and hours of phone calls, the visits from company executives, the "Boeing Contribution," and all the rides in M-Cab—had failed to persuade the safety board.

> Analysis of the human performance data, including operational factors, does not support a scenario in which the flight crew of USAir Flight 427 applied and held a full left rudder input until ground impact more than 20 seconds later.

Hall switched gears and read several conclusions about the Colorado Springs crash. It was an extraordinary moment. The board was reopening the eight-year-old investigation and declaring that it had solved the mystery.

> Analysis of the CVR, Safety Board computer simulation, and human performance data, including operational factors, from the United Flight 585 accident shows that they were consistent with a rudder reversal.

When Hall got to the conclusion that blamed the Colorado Springs crash on a reversal, the families broke into applause. But Hall admonished them, "Please, no demonstrations."

He then read the probable cause statement for the USAir crash—four and half years of investigation boiled down to two sentences:

> The probable cause of the USAir Flight 427 accident was a loss of control of the airplane resulting from the movement of the rudder surface to its blowdown limit. The rudder surface most likely deflected in a direction opposite to that commanded by the pilots as a result of a jam of the main rudder power control unit servo valve secondary slide to the servo valve housing offset from its neutral position and overtravel of the primary slide.

Translation: The rudder probably reversed. The fault was with the airplane, not the pilots.

"Any comment or discussion?" Hall asked, but no one replied. "If not, do I hear a motion that the findings and probable cause be adopted?"

Hammerschmidt made the motion and Francis made the second.

"Seconded," Hall said. "All in favor, please signify by saying 'aye.'"

"Aye."

"Aye."

"Aye."

"Aye."

"The findings and probable cause are unanimously adopted by the National Transportation Safety Board," Hall said.

Behind him, Smith grinned at Daniels and raised her eyebrows in relief.

In the audience, some family members wept. Others choked back their tears.

Hall invited the board members each to make a final statement. Francis spoke of the importance of flight data recorders. Hammerschmidt said the investigation provided valuable research on wake turbulence. Black said it had been a frustrating experience.

"We engineers normally like to base our decisions on hard, cold facts— measurable things, things we can lay our hands on," he said. "And while there is much evidence in these accidents, the vast majority of it is not hard, cold evidence." He said he and his family often flew 737s and that the plane had "a good, documented safe history. There's an awful lot of successful flights out there."

As Black left the stage, he told reporters he had misgivings about the case. "I damn near voted against it," he said. "This is a circumstantial case."

Hall, renowned for blasting the FAA and Boeing, was uncharacteristically muted. He told reporters that he flew 737s every week, an endorsement of their safety. His only strong comments dealt with the FAA's sluggishness on flight data recorders. He said little about the rudder. (He later said he was subdued because he did not want to anger the FAA before it began the 737 engineering study.)

After the meeting, Boeing officials held a news conference in the only meeting room they could find in the small hotel—the bar. The company took a conciliatory approach on the 737 report, and no longer pushed the pilot-error theory.

"We respect the board's opinion," said Charlie Higgins, Boeing's vice president for airplane safety. He said new rudder valves being installed in 737s "completely eliminate any possibility of a reversal."

The company will "do everything we can to look at the 737 rudder system and see if there is anything that can be improved," he said. Higgins looked to the back of the room where several family members stood, and acknowledged their loss.

"I'd like to sincerely offer our condolences to the families," he said. "It's small consolation to them, but I believe this accident has improved aviation."

Out in the hallway Brett said he felt vindicated.

"Pretty hard-hitting," he said to his lawyer, Mike Demetrio.

"I agree," Demetrio said. "They took the approach that the plane was deficient from Day One."

Demetrio said some people were worried that the board would cave to pressure from Boeing, but that didn't happen. "Today, I think the taxpayers got their money's worth from the federal government," he said.

Brett stepped inside the temporary NTSB office across from the ballroom and thanked several of the investigators. He told Loeb, "I really appreciate your being vocal yesterday."

"That's what they pay me to do," Loeb said.

A few minutes later Brett said he was pessimistic that Tom McSweeny, the FAA official in charge of airplane certification, would agree with the NTSB's recommendations. Brett and other family members had met with McSweeny that morning, and Brett came away with the impression that the FAA official was more interested in making excuses than in being aggressive about the rudder problem.

Brett said he regretted that the NTSB had been unable to solve the Colorado Springs mystery eight years earlier. If they had solved it, he said, "Joan would be alive today."

Upstairs in the FAA office, McSweeny stubbornly refused to give the NTSB credit. He said his agency would take further action about the 737, but he said it was not because of anything the NTSB had done. He said the FAA would call for faster minimum speeds to ensure that 737s did not fly slower than the crossover point, but he insisted, "We're doing that because of the unexplained MetroJet incident. It is not a response to this meeting."

When Haueter returned to the NTSB office at L'Enfant Plaza, everyone was glum. They were angry about Black's and Hall's comments. The board members were backing away from the report they had just approved. "The report was adopted unscathed, but it felt like cold water was thrown on it," Haueter said.

To perk up the troops, he went downstairs to the L'Enfant Plaza mall and bought four bottles of champagne and some plastic glasses. He summoned the Flight 427 team to the conference room.

They reminisced about the investigation. Malcolm Brenner told the group that Haueter must be blessed with some kind of superhuman hormone that kept him going when others were ready to quit. Someone else proposed a toast to Eastwind pilot Brian Bishop, who had survived a rudder reversal, saved fifty-five lives, and given the NTSB key evidence about what was wrong in the 737.

"To Captain Bishop," one of the investigators said, and they all clinked their plastic cups together.

EPILOGUE

With the NTSB investigation complete, lawyers for USAir, Boeing, and the families quickened the pace of their settlement talks. The companies were eager to settle with all families before the Cook County trial, scheduled to begin in November 1999.

Two years earlier, USAir's insurance company had offered Brett $2 million to settle. The company told Demetrio that it was a reasonable offer because that's what the company had paid other people in similar circumstances. Brett didn't think $2 million was enough. He said he wanted to go to court.

To reach a settlement amount for Joan, both sides had hired economists and accountants. They had estimated her lifetime earnings if she had not been killed in the crash, as well as the cost for Brett to hire someone else to perform her household chores. The economists then subtracted the "terminated consumption"—the amount Joan would have spent each year on clothes, food, and so on. The calculations were all part of the painful but necessary process of setting a price on Joan's life.

There was no agreement on the numbers. The economist hired by Brett's lawyers estimated his total loss from Joan's death at $1.6 million to $1.8 million. An economist hired by USAir estimated the loss at $1.4 million, while one hired by Boeing came in at $833,000.

One reason for the disparity was that the parties disagreed on how long Joan would have worked before retiring. The economist hired by Brett's firm assumed 31 years after the crash, USAir used 29.5 years, and Boeing assumed

28.5. Also, Brett's economist came up with a higher value for Joan's household chores. For 1999, for example, he said the chores were worth $10,974. USAir's estimate was $9,993, while Boeing's was $7,011.

Why such a range? The economists used different sources for their calculations and applied various factors to adjust them for the future. Their assumptions usually benefited their client's interests. In general, the economist used by Brett's lawyer relied on more generous assumptions, while the USAir and Boeing economists used more conservative figures.

The numbers provided a starting point for the settlement talks. In 1998, a year after Brett rejected the $2 million offer, his lawyer, Mike Demetrio, had made a counteroffer to USAir and Boeing, asking for $5 million. Demetrio said that figure reflected the $1.8 million estimate from his economist, plus a sizable amount for Brett's "loss of society," the legal term for the loss of love and companionship when a spouse dies. There also was a "significant premium" to account for the long delay since the 1994 crash, Demetrio said.

But USAir and Boeing couldn't respond because the companies were too busy fighting each other about how much each should pay. At some depositions, the Boeing and USAir lawyers were "screaming at each other at the top of their lungs," Demetrio said. "It was like watching my third-grader on the playground."

In February 1999—about one month before the NTSB meeting—Boeing and USAir lawyers put aside their feud and floated the possibility of a $3 million settlement. But Brett again said it was too low. The case was headed for trial.

He viewed the trial as his revenge against the companies. He said he wanted "to stick it to them financially" and embarrass them in court. The USAir and Boeing executives would get a public grilling from Demetrio and other lawyers. For weeks, the companies would be battered by the bad publicity.

As the trial date neared, both sides knew there would be last-minute settlement talks. That was the nature of crash lawsuits. The parties often settled only a few hours before the trial was to start. There were five cases left in Cook County, including Brett's. They were scheduled to be tried together.

The lawyers had put together powerful exhibits to sway the jury and create public relations problems for Boeing and USAir. The biggest was a full-size, fully functional 737 tail. It was three stories high and painted in USAir's colors. The lawyers planned to put it outside the courthouse and demonstrate a rudder reversal in full view of the public and the news media. For Boeing and USAir, that would be a public relations nightmare.

Demetrio and the other plaintiff's attorneys had another powerful prop: a four-foot-long replica of a USAir 737 with a removable top. Demetrio was going to lift the top off and show jurors the inside. He planned to say: "This was where Joan Van Bortel was sitting."

Negotiations began Monday, November 1, two days before the trial was supposed to start. By the evening of November 2, USAir and Boeing had come up with a new offer for Brett: $6 million. It was double their last offer and $1 million more than Brett's request a year earlier.

Demetrio called his client shortly before 5 P.M. "We think it's a good offer," he said.

"Do I have time to think about this?" Brett asked.

Demetrio said he needed an answer that night.

Sitting in his office in Oak Brook, a Chicago suburb, Brett pondered his options. He had wanted a trial to publicly embarrass Boeing and USAir and force them to be more responsible with safety issues in the future. He felt that Boeing was largely responsible for the crash, that the company knew about the rudder problem after the Colorado Springs accident but did nothing to correct it. Brett believed the company had rolled the dice with people's lives rather than paying to fix the planes. He also blamed USAir for allowing the plane to have dirty hydraulic fluid, which he believed had contributed to the crash.

On the other hand, Brett wanted to start a foundation named after Joan to pay for scholarships. The money would allow him to start it immediately. Six million dollars was triple the original offer, and Demetrio said it was high for a victim with no children. If Brett went to trial, he would be taking a big risk. He might get less money, and he might not see it for years because of appeals.

Or the trial might give him a lot more.

Time was of the essence. Joan's father had died recently, and her mother was getting older. Brett wanted Joan's mother to know that the scholarship foundation was going to be a reality. So he called her at home in Melrose, Iowa, and asked her opinion. She said she was comfortable with Brett making the decision.

After he hung up, Brett came up with a novel way to decide.

Fate had put Joan on Flight 427, so it might as well be fate that decided whether he should settle. He would flip a coin. Maybe God or Joan's hand would decide which way it landed.

He reached into a drawer where he kept his spare change and grabbed a quarter. It was from 1987, which seemed appropriate because that was the year he and Joan had started dating. Heads he would go to trial. Tails he would settle.

He flipped the coin and was sure it would come up heads. As it tumbled through the air, he thought about the trial and what it would be like. The quarter landed on the carpet.

Tails.

It was time to settle.

He took a long walk and then called Demetrio from his cell phone. He wanted to be sure a settlement wouldn't prevent him from writing a book about his experiences. Would there be a gag order?

No, Demetrio assured him. Brett was free to talk.

"Let's end it," Brett said.

USAir Flight 427 changed aviation forever.

At the NTSB's urging, the FAA conducted an unprecedented study of the 737 rudder system. An independent panel of hydraulic engineers and flight control experts spent a year studying the valve, flying 737 simulators and analyzing extreme failures that were not fully explored in the Flight 427 investigation.

In its final report, the panel agreed with the NTSB and warned that the 737 rudder system was susceptible to many failures and jams that could be catastrophic. The group recommended better training for 737 pilots and, in the long term, a complete redesign of the plane's rudder system.

As a result of the report, the FAA and Boeing announced in September 2000 that the unique valve-within-a-valve rudder system on all 737s would be replaced by a two-valve system, similar to the one used on other planes. The announcement marked a surprising and dramatic change for Boeing and the FAA. Both had insisted for years that the redesign was unnecessary.

Boeing said it would pay the entire cost of the new rudder system, estimated to be more than $240 million. The company said the new system should be installed in all 737s by 2007.

In the meantime, all 737s have been equipped with improved valves that cannot reverse. The planes also have better yaw damper couplers (the computers that command small adjustments to the rudder) and pressure limiters (devices that limit how far a rudder will move). Pilots have been alerted to the crossover point and have been trained to identify and recover from rudder problems.

After years of complaints from the NTSB, the FAA mandated that airlines upgrade flight data recorders to take additional measurements such as rudder pedal position, a change that should make it easier for investigators to determine the cause of future crashes. Instead of the thirteen parameters recorded on the USAir plane, they now must have at least seventeen.

Tom McSweeny, the FAA official who was criticized for being soft on Boeing, left the FAA in the fall of 2001. He is now director of international safety and regulatory affairs for Boeing.

USAir changed its name to US Airways in early 1997 after former United Airlines head Stephen Wolf became chairman. One of Wolf's first priorities was a new image that has helped erase the memories of the airline's five crashes. US Airways repainted its fleet, announced plans to buy new Airbus planes (the airline said the 737 problems were not a factor in the decision), and improved amenities for frequent travelers. A plan to merge with United Airlines fizzled in 2001, so USAir once again talked about consolidating and cutting costs. Its future was unclear as this book went to press.

Boeing merged with its rival McDonnell Douglas in 1997. One of the biggest challenges for the gargantuan company has been keeping up with demand. Sales of 737s remained strong and were unaffected by the controversy over the rudder. In 2000 Boeing produced more than twice as many 737s as in 1994, the year of the crash.

The painful experiences of the families of Flight 427 and other crashes prompted Congress to pass the Aviation Disaster Family Assistance Act, which requires airlines to have detailed plans for responding to a crash and notifying family members. An airline must notify the next of kin as soon as it verifies that the passenger was aboard the plane—regardless of whether other names have been confirmed, and the airline must consult with families about plans for human remains and personal effects. Also, the law says attorneys cannot make unsolicited contact with families until at least thirty days after an accident.

Under the new law, the NTSB is designated as the main federal agency to help families after an aviation disaster. Many safety board investigators, including Haueter, did not want their agency to take on that responsibility because they believed it conflicted with the safety board's investigative mission. But Chairman Jim Hall said it was consistent with the government's role to help people in need. Since the passage of the act, airlines and the NTSB have been praised for their treatment of families.

In late 2000, Tom Haueter was promoted to deputy director of aviation safety, the second-highest aviation job in the NTSB. He says he is confident that his team solved the mystery of Flight 427, although it's possible that the 737 rudder system may have other failure modes that haven't been found. "The big lesson is to keep pushing," he says. "In 585 [the Colorado Springs case], we ran out of data and quit. In 427, we didn't quit."

In the summer of 2001 the NTSB released a revised report on United Airlines Flight 585, the Colorado Springs crash. It incorporated the findings of the Flight 427 investigation and the new conclusion that the United plane most likely had a rudder reversal. The words "FOR UNDETERMINED REASONS" no longer appear on the cover.

Haueter still flies in 737s on business trips, but his constant talk of rudder problems has frightened his wife, Trisha, to the point that when they go on vacation, she insists on flying in a different type of plane.

Jean McGrew retired from Boeing in 2000. He says the crash investigation took a heavy toll on him and was a big factor in his decision to retire. Always blunt about his feelings, he says he is still convinced the pilots caused the crash.

However, McGrew says he is glad the 737's unique rudder valve is being replaced. "It was not as good as it should have been," he says. "It could have been more fail-safe than it was."

He has strong feelings about the NTSB and how it was run by politicians: "I think they ought to take the politicians and get rid of them."

John Cox was promoted to fly the Airbus A320 for US Airways and in 2001 was named chairman of ALPA's Executive Air Safety Committee, the union's top safety job. During the USAir investigation, a fellow pilot sent him a book with this inscription: "John, there are people alive today that otherwise would not be except for your work in safety, and I thank you for that."

On September 9, 2000, the day after the sixth anniversary of the crash, Brett Van Bortel got engaged to Victoria Hartz, a woman he had been dating for several years. They were married in July 2001.

His lawyers received $1.5 million, which was 25 percent of his settle-ment. After other expenses were deducted, Brett ended up with $4.18 mil-lion. He is using $1 million for the Joan Lahart-Van Bortel Memorial Schol-arship Fund. Each year, it provides a four-year scholarship for a young woman from Joan's home county in Iowa to attend college. Brett also used some of his settlement to take Joan's mother and brother to Ireland.

Brett says he has healed as well as anyone can. Coping with the crash "is something that never gets better, it just recedes further in the past."

As I completed interviews for this book in 2001, many people at the NTSB wanted to take credit for the success of the Flight 427 investigation. There were so many conflicting claims that it took me several days to sort them out. Ultimately, I decided it doesn't matter who came up with the idea for the independent engineering panel or who devised the phrase "reliably redun-dant." The truth is that the Flight 427 investigation was an extraordinary team effort. If it weren't for the odd mix of characters and brilliance and pure good luck, the case might still be open.

Haueter's quiet persistence kept the investigation plodding along when some at the safety board wanted him to give up. His friendly, low-key style allowed a healthy exchange of ideas. His friend Greg Phillips, the systems group leader, was an important voice of caution. Phillips kept the NTSB from rushing to judgment and blaming the valve when there was insufficient proof.

Loeb, the aggressive manager, provided an important spark when the investigation reached its lowest point. He cleared the way for the first batch of safety recommendations and became the forceful advocate Haueter needed at the end.

Chairman Hall's lack of technical experience and his country bumpkin persona often frustrated the investigators. But his outsider perspective led to the most important breakthrough. It was Hall's idea to create a panel of the Greatest Minds in Hydraulics, a suggestion that some investigators—includ-ing Haueter—opposed. The panel proposed the thermal shock test, which became the turning point of the investigation.

Board members George Black and Bob Francis also provided an impor-tant perspective, toning down a report that was stronger than the facts justi-fied. They made sure it was based on solid ground.

The NTSB party system has often been criticized for supposedly allowing companies to manipulate an investigation. But in this case, it worked marvel-ously. The key discovery in the investigation—that the valve could reverse—was made not by someone at the NTSB but by an engineer at Boeing. The com-pany provided immeasurable technical experience to the probe, as well as mil-lions of dollars in tests that the tiny safety board could never have afforded. The constant fighting between Boeing and ALPA crystallized the issues for the safety board and ensured that every theory was considered.

During the six years I worked on this book, I heard a lot of criticism of the NTSB. I approached the investigation with an open mind. I considered the possibility that the party system was flawed or that the NTSB was a puppet that could be controlled by a powerful corporation such as Boeing. But I found just the opposite. The party system led to important breakthroughs, and Haueter and his investigators were not cowed by the mighty Boeing Company. I also believe that the NTSB's messy internal fights led to a more solid report and that the tension between the NTSB and the FAA creates a healthy check and balance.

In the case of Flight 427, the system worked.

GLOSSARY

ailerons   Panels on the wings that control the roll of the plane and allow it to bank to the right or left.

ALPA   The Air Line Pilots Association, a labor union.

control column   The "stick" in the cockpit that pilots use to make the plane climb and descend. On top of the control column is the wheel, which is used to control the roll of the plane.

crossover point   The critical airspeed at which a full swing by the 737's rudder cannot be counteracted by the ailerons. When a plane has a rudder hardover while flying slower than the crossover point, the pilot must speed up to regain control.

CVR   Cockpit voice recorder. Also known as a "black box," it records sounds and pilot conversations that help investigators determine what caused a crash.

flaps   Movable panels on the wings that provide extra lift for a plane at slower airspeeds.

fly-by-wire   A computerized system in newer airplanes that sends electronic signals to move the flight controls. The Boeing 737 is not a fly-by-wire plane. It has cables that move back and forth to send commands to the flight controls.

hardover A malfunction that occurs when the rudder or another flight control suddenly moves as far as it can, usually because of a problem with a hydraulic device.

jump seat A fold-down seat in the cockpit that allows an FAA inspector or a company official to observe the pilots. Most airlines allow their pilots to ride in the jump seat to commute from their home city to their crew base.

M-Cab The special Boeing flight simulator in Seattle that was used to re-create the crash and test scenarios about what happened.

party system The NTSB practice of allowing companies and labor unions to take part in an investigation.

power control unit (PCU) The hydraulic device that moves the rudder or another flight control. The rudder PCU on the 737 is about the size of an up-right vacuum cleaner.

rudder The movable vertical panel on the tail. On the 737, pilots use it primarily when landing in a strong crosswind or on the rare occasion when they have an engine failure.

servo valve A soda can–size valve inside the PCU. The 737 has a unique version known as a dual concentric servo valve. It has two tubes that slide back and forth. The slides send bursts of hydraulic fluid against a piston that moves the rudder.

stall The result when a plane no longer has enough air moving over its wings to stay aloft.

stickshaker A device that rattles the pilots' control columns to warn them that the plane is about to stall.

wheel The steering wheel–like device in the cockpit that pilots use to move the ailerons and flight spoilers, the panels on the wings that cause the plane to roll to the left or right. The wheel is on top of the control column.

windscreen The front window on an airplane, like a windshield on a car.

yaw damper A device that creates a smoother ride by making hundreds of small adjustments to the rudder during a flight.

SOURCES

This book is based on hundreds of hours of interviews that I conducted over six years with the principal characters—Tom Haueter, Brett Van Bortel, John Cox, and Jean McGrew. I spent many Saturdays in Haueter's living room and went flying with him in his Stearman. Likewise, I spent many afternoons in John Cox's home in St. Petersburg, Florida, where he taught me about airplane systems, rudder valves, and crash investigations. He persuaded USAir to allow me to ride with him in the cockpit for a four-day trip in January 1997, a trip that helped me better understand the life of a pilot. We had a grueling overnight in Boston and one that was not so grueling in San Juan, Puerto Rico.

I traveled to Seattle three times to interview McGrew and spoke with him by phone on several other occasions. A Boeing public relations official was present for most of those interviews, but McGrew was still candid in expressing his feelings about the NTSB.

I interviewed Brett Van Bortel in Chicago four times, visited the crash site with him on the first anniversary of the crash, and was with him during the final meeting in Springfield, Virginia. We exchanged E-mail frequently and spoke often by phone.

I also interviewed emergency workers in Hopewell Township, FAA officials, Brett's friends and relatives, the NTSB investigators on Haueter's team, the NTSB board members (except for John Hammerschmidt, who declined to be interviewed), USAir employees, Boeing engineers and test pilots, and Roxie Laybourne, the Smithsonian feather expert. (See list below.)

I was present for many of the scenes in the book, including both public hearings, Chairman Jim Hall's ride in the M-Cab simulator, a portion of the "fat guy" tests, and the final meeting in March 1999. I was able to write about several other scenes—including Joan's memorial service, the first anniversary ceremony, and the recovery of passenger belongings from the trash bins in March 1995—thanks to videotapes that were recorded by family members or friends.

Other scenes are reconstructed based on the recollections of the main characters. When possible, I verified their accounts with others. During that process, I discovered that Haueter, Cox, and Brett each had a remarkably accurate memory and a keen eye for detail. In addition to the interviews, I relied heavily on transcripts of the public hearings and the NTSB's huge docket on the crash, which includes more than 10,000 pages of investigative reports and documents from USAir and Boeing. Depositions from the Chicago court file provided details about the scene aboard Ship 513 when Andrew McKenna heard the gurgling sound and the scene at O'Hare before the plane departed for Pittsburgh.

I relied on the excellent coverage in the *Pittsburgh Post-Gazette* for details about the days immediately following the crash and on the *Palm Beach Post* for the profile of Paul Olson, the convicted drug dealer on the plane. I also relied on the *New York Times* coverage of Orville Wright's crash in 1908 and the Knute Rockne crash in 1931.

The section on Boeing's history and the development of the 737 is based on old memos in the company's archives, and on two superb books: *Legend and Legacy*, by Robert Serling, and *Flying High*, by Eugene Rogers.

Interviews

The National Transportation Safety Board (NTSB)

Tom Haueter, investigator-in-charge
    Trisha Dedik, Tom Haueter's wife
Greg Phillips, systems group chairman
Jim Cash, cockpit voice recorder analyst
Malcolm Brenner, human factors specialist
Tom Jacky, aircraft performance specialist

NTSB managers

Bud Laynor, deputy director of aviation safety (until 1996)
Bernie Loeb, director of aviation safety, 1996–2001
Ron Schleede, deputy director of aviation safety, 1996–1999
Peter Goelz, managing director
John Clark, director of aviation safety, 2001–

Board members

Chairman Jim Hall
    Deb Smith, Jim Hall's assistant
John Goglia
George Black
Robert Francis
    Denise Daniels, Robert Francis's special counsel

The Air Line Pilots Association (ALPA)

John Cox, systems group member and USAir pilot
Herb LeGrow, coordinator for USAir 427 investigation
Keith Hagy, manager of accident investigation
Bill Sorbie, chairman of central air safety committee/USAir

The Boeing Company

Jean McGrew, chief engineer for the 737
John Purvis, chief of air safety investigations
Rick Howes, air safety investigator and coordinator for USAir 427
Michael Hewett, flight test pilot
Michael Carriker, flight test pilot
Ed Kikta, hydraulics engineer
Jim Draxler, systems engineer and Ed Kikta's boss
Jim Kerrigan, senior principal engineer
Mike Denton, 737 chief engineer (1997)
Jack Steiner, vice president (retired)
Martin Ingham, specialist engineer
Ragnar Nordvik, regional marketing director
Paul Martin, senior instructor pilot

USAir

Ralph C. Miller, manager of the Next-of-Kin Room
Deborah Thompson, director of consumer affairs
Dave Supplee, mechanic and accident investigator for the machinists union

Federal Aviation Administration (FAA)

Vikki Anderson, accident investigator
Thomas McSweeny, director of aircraft certification
Steve O'Neal, flight test engineer
Ed Kittel, bomb expert
Ken Frey, systems engineer

Sharon Battle, operations center officer
David Canoles, director air traffic effectiveness
David Thomas, director of accident investigation
Bud Donner, manager of accident investigation
Les Berven, engineering test pilot
Dick Paul, engineering test pilot

Others

Brett Van Bortel
Bonnie Van Bortel, Brett's mother
Mike Demetrio, Brett's attorney
Brian Bishop, Eastwind Airlines pilot
Federico Pena, Secretary of Transportation
Wayne Tatalovich, Beaver County coroner
Russ Chiodo, Beaver County director of emergency operations
Captain James Rock, Hopewell Township volunteer firefighter
Fred David, Hopewell Township police chief
George David, owner of farm adjacent to crash site
Rudy Kapustin, former NTSB investigator
Nancy Edwards, coworker of Joan Van Bortel
Jen Brundage, friend of Joan Van Bortel
David Hause, deputy medical examiner, Armed Forces Institute of Pathology
Roxie Laybourne, feather expert, Smithsonian Institution
John Little, assistant security manager, Museum of Flight
John Kretz, executive director, Flight 427 Air Disaster Support League
Don Hunt, professor, Embry-Riddle Aeronautical University
Michael Pangia, attorney
Ralph Vick, member of Expert Panel on Hydraulics

Articles and Books

Acohido, Byron. "Safety at Issue: The 737." *Seattle Times,* reprint of Parts 1–5, October 27–31, 1996.
Associated Press. "USAir Passengers Cited Noise." *Philadelphia Inquirer,* November 21, 1994.
Bayles, Fred, and Robert Davis. "Doubts Shadow Flight 427 'Party' Inquiry." *USA Today,* March 23, 1999.
Bean, Ed. "Damage Control: After 137 People Died in Its Texas Jet Crash, Delta Helped Families." *Wall Street Journal,* November 7, 1986.
Belden, Tom. "On the Ropes, USAir Takes the Offensive." *Philadelphia Inquirer,* November 27, 1994.
———. "Timing Is Bad for an Airline in a Tailspin." *Philadelphia Inquirer,* September 10, 1994.
Belko, Mark. "Crash Photos Available to Families." *Pittsburgh Post-Gazette,* January 11, 1995.

———. "Mission of Mercy." *Pittsburgh Post-Gazette,* October 16, 1994.

———. "Morgue Workers' Somber Job at an End." *Pittsburgh Post-Gazette,* October 8, 1994.

Brelis, Matthew. "In Pa., Seeking Clues amid Airliner's Wreckage." *Boston Globe,* September 10, 1994.

Bryant, Adam. "USAir Crash Investigators Rule Out More Causes." *New York Times,* September 16, 1994.

———. "With Factors Ruled Out, USAir Crash Emerges as Puzzle." *New York Times,* September 17, 1994.

"Conference Is Called on Rockne Air Crash." *New York Times,* April 18, 1931.

Cooke, Patrick. "When Airplanes Crash, Flight Recorders Often Tell Why." *Air and Space,* July 1988, 31–37.

Creedy, Steve. "Audit Deems USAir Safe, Revises Procedures." *Pittsburgh Post-Gazette,* March 18, 1995.

———. "Schofield Retiring as Chief of USAir." *Pittsburgh Post-Gazette,* September 7, 1995.

Cushman, John H., Jr. "Crash Investigators Broaden Their Inquiry." *New York Times,* September 13, 1994.

Dockser, Amy. "Infidelity Becomes Factor in Wrongful-Death Suits." *Wall Street Journal,* August 10, 1989.

Editorial. "Ice in the Air." *New York Times,* April 9, 1931.

Ellicott, Val. "Air Crash Victim Had Built New Life." *Palm Beach (Fla.) Post,* September 13, 1994.

"Fatal Fall of Wright Airship." *New York Times,* September 18, 1908.

"Figures Show Flying Is Safer Than Standing Behind a Mule." *New York Times,* January 3, 1927.

Frantz, Douglas, and Ralph Blumenthal. "Troubles at USAir: Coincidence or More?" *New York Times,* November 13, 1994.

Galvin, Thomas. "USAir May Nose-Dive After New Safety Shocker: Experts." *New York Post,* November 14, 1994.

"Government Bans 35 Fokker Planes." *New York Times,* May 3, 1931.

Haddigan, Michael. "Drug Informant Was on USAir Plane That Crashed." Associated Press, September 15, 1994.

———. "No Plot Against Man Is Found in Fatal Crash." *Philadelphia Inquirer,* September 16, 1994.

Hanchette, John. "Drug Defendant Not Linked to USAir Crash, Lawyer Says." Gannett News Service, September 16, 1994.

———. "Witness on Doomed Plane Tied to Colombian Drug Trade." Gannett News Service, September 15, 1994.

Harr, Jonathan. "The Crash Detectives." *New Yorker,* August 5, 1996, 34–55.

"Jet Rolled, Plummeted." *Pittsburgh Post-Gazette,* September 10, 1994.

Knowles, Robert G. "USAir May Face Wary Aviation Market." *National Underwriter,* September 19, 1994.

"Knute Rockne Dies with Seven Others in Mail Plane Dive." *New York Times,* April 1, 1931.

Kohn, Bernie. "Latest Accident Comes at Key Time for Carrier." *Charlotte (N.C.) Observer,* September 9, 1994.

———. "USAir Stresses Safety, Admits Trust a Challenge." *Charlotte (N.C.) Observer,* September 10, 1994.

Kolata, Gina. "When Is a Coincidence Too Bad to Be True?" *New York Times,* September 11, 1994.

Komons, Nick A. *Bonfires to Beacons: Federal Civil Aviation Policy Under the Air Commerce Act, 1926–1938.* Washington, D.C.: U.S. Department of Transportation, 1978.

Langewiesche, Wolfgang. *Stick and Rudder.* New York: McGraw-Hill, 1944.

McDowell, Edwin. "737 Warning Draws a Shrug from Public." *New York Times,* February 22, 1997.

McLeod, Douglas. "USAir Crash Should Not Lead to Aviation Rate Hikes." *Business Insurance,* September 12, 1994.

Mossman, John. "Clues Sought in Flight 585 Crash That Claimed 25 Lives." Associated Press, March 4, 1991.

"Orville Wright Is Not So Well." *New York Times,* September 20, 1908.

Oster, Clinton V., Jr., John S. Strong, and C. Kurt Zorn. *Why Airplanes Crash: Aviation Safety in a Changing World.* New York: Oxford University Press, 1992.

Patterson, James. "The Drug Dealer on a Fatal Crash." *Indianapolis (Ind.) Star,* December 17, 1994.

Petzinger, Thomas, Jr. *Hard Landing: The Epic Contest for Power and Profits That Plunged the Airlines into Chaos.* New York: Times Business/Random House, 1995.

Phillips, Don. "Panel to Seek Redesign of 737 Rudder." *Washington Post,* April 16, 2000.

———. "Probe of Two Deadly Crashes Points to Rudders." *Washington Post,* March 25, 1999.

Pomfret, John D. "Photos Said to Show Pilots Ignoring Duty." *New York Times,* September 7, 1962.

Reyes, David, and Nancy Wride. "Anxious Relatives Frustrated at Lack of Crash Information." *Los Angeles Times,* August 17, 1987, Metro.

"Rockne Air Crash Is Laid to Ice-Covered Wings." *New York Times,* April 8, 1931.

"Rockne Air Crash Laid to Lost Wing." *New York Times,* April 3, 1931.

Roddy, Dennis. "Officials Have First Hints on Why Flight 427 Went Down." *Pittsburgh Post-Gazette,* September 11, 1994.

Rodgers, Eugene. *Flying High: The Story of Boeing and the Rise of the Jetliner Industry.* New York: Atlantic Monthly Press, 1996.

Rodgers-Melnick, Ann, and Rebekah Scott. "Finding Solace in the Church Community." *Pittsburgh Post-Gazette,* September 12, 1994.

Sammon, Bill. "Pittsburgh Stops to Mourn." *Cleveland Plain Dealer,* September 13, 1994.

Serling, Robert J. *Legend and Legacy: The Story of Boeing and Its People.* New York: St. Martin's Press, 1992.

Shapiro, Stacy. "Aviation Market More Selective." *Business Insurance,* October 31, 1994.

Smith, Matthew P. "Wake Still Top Suspect in Crash." *Pittsburgh Post-Gazette,* October 9, 1994.

Steigerwald, Bill. "It Still Hurts; Survivors Group Works to Improve Airline Treatment." *Pittsburgh Post-Gazette,* September 8, 1995.

Stewart, Stanley. *Flying the Big Jets.* 3rd ed. Stillwater, Minn.: Specialty Press, 1992.

Taylor, Alex, III. "Boeing Sleepy in Seattle." *Fortune,* August 7, 1995, 92–98.

Twedt, Steve. "Jet's Right Engine Parts Studied." *Pittsburgh Post-Gazette,* September 12, 1994.

"Unable to Fathom Rockne Plane Crash." *New York Times,* April 2, 1931.

Underwood, Anne, and Melinda Beck. "Unsolved Mystery." *Newsweek,* September 11, 1995, 48–52.

United Press International. "Stewardesses Tell House Unit Pilots Let Them Fly Airliners." *New York Times,* September 29, 1962.

Walsh, Lawrence. "Families of Flight 427 Victims Give Human Voice to Dry NTSB Inquiry." *Pittsburgh Post-Gazette,* January 29, 1995.

———. "Sincerity Not Long Suit of Crash Suit Seekers." *Pittsburgh Post-Gazette,* October 2, 1994.

———. "USAir Will Add Crash Notation to Monument." *Pittsburgh Post-Gazette,* March 1, 1995.

Weir, Andrew. "Secrets of the Black Box." *Reader's Digest,* July 1997, 75–80.

"Wilbur Wright Shocked." *New York Times,* September 19, 1908.

Wilson, John R. M. *Turbulence Aloft: The Civil Aeronautics Administration amid Wars and Rumors of Wars, 1938–1953.* Washington, D.C.: U.S. Department of Transportation, 1979.

Wooton, Suzanne. "Like a Death in the Family." *Baltimore Sun,* September 18, 1994.

ACKNOWLEDGMENTS

Tom Haueter, John Cox, and Brett Van Bortel made this book possible because of their candor. For six years, they shared their private thoughts about the investigation and how the crash had changed their lives. They endured my frequent visits, telephone calls, and repetitive questions. I thank them for their patience and their willingness to open their lives to thousands of readers. I also appreciate the help I received from Trisha Dedik and Jean Cox, who discussed how the investigation affected their husbands.

I am grateful for the cooperation of the chairman of the NTSB, Jim Hall, and the managing director, Peter Goelz. The agency's public affairs office initially rejected my proposal for a behind-the-scenes look at the investigation, but Hall and Goelz agreed because they believed they had a positive story to tell. They granted me special access to the investigators on the condition that I not publish anything until the report was complete. I also want to thank the NTSB public affairs staff, past and present, including Mike Benson, Pat Cariseo, Ted Lopatkiewicz, and Alan Pollock.

I appreciate the openness of people at Boeing. The company had never cooperated with a project like mine, but several key officials realized it was in Boeing's interest—and the interest of passengers who fly its planes—to tell its side. I appreciate the support of Bill Curry, Liz Verdier, Russ Young, Sue Bradley, John Dern, and Steve Thieme. Boeing's historian, Tom Lubbesmeyer, shared the company's memos and marketing materials from the 1960s, which provided tremendous insight into the decision-making process when the 737 was designed. The Boeing engineers and pilots involved in the investigation—Jean McGrew, John Purvis, Rick Howes, Mike Hewett, Mike Carriker, and Jim Draxler—were honest about their feelings and frustrations about the NTSB. Because of their candor, I was able to write a more balanced book that reveals the tensions and disagreements of the investigation.

I am thankful for the assistance from people at USAir and the Federal Aviation Administration. At USAir: Rick Weintraub, Deborah Thompson, Dave Supplee, George Snyder, and Ralph Miller. At the FAA: Vikki Anderson, Dave Thomas, Drucella Andersen, Bud Donner, Ed Kittel, Eliot Brenner, Diane Spitaliere, Paul Turk, Bob Hawk, and Ned Preston.

Thanks also to Joe Formoso, Mike Demetrio, Tom Ellis, Michael Pangia, Russ Chiodo, John Kretz, Steve Okun, John Masor, Bob Flocke, and Keith Hagy.

I am deeply grateful to my colleagues at the *St. Petersburg Times* who helped with my series *28 Seconds,* on which this book is based, and to the *Times* for waiving copyright on the material first published in the series. I am indebted to Richard Bockman and Neil Brown, who helped me shape the early drafts and provided crucial advice on how to tell such a complex story. Thanks also to Paul Tash, Sara Fritz, Chris Lavin, Susan Taylor Martin, Kelly Boring Smith, Bill Serne, Tom Rawlins, David Dahl, Sherry Robinson, Kitty Bennett, and *Times* attorneys George Rahdert and Allison Steele. I also thank my friend Don Phillips of the *Washington Post,* who provided help and encouragement along the way.

I am indebted to people who read drafts of the manuscript at various stages, including Pat Trenner, Eric Adams, Peter Wallsten, Scott Moyers, and John Donnelly. My agents, David Black and Gary Morris of the David Black Literary Agency, provided tremendous support during the ups and downs of the past six years. I am especially grateful to Mark Gatlin at the Smithsonian Institution Press for his enthusiasm and persistence about the project.

I thank my in-laws, Frank and Otey Swoboda, for providing me a place to write. My wife, Katherine, provided many valuable suggestions about the manuscript, and she and our children, Molly, Annie, and Miles, tolerated my frequent trips and my six-day workweeks as I finished the book. We'll have time to play with the Sega Dreamcast now, guys.

INDEX

United Airlines, 99, 208
United Airlines Flight 585 crash (1991), 36,
93–96, 138, 169, 202, 204; analysis of
rudder system, 31, 95, 96; damage to
rudder system, 69, 71; original NTSB
report, 96, 195; revised NTSB report, 209;
rudder reversal, 202, 209; wind, 95, 96,
193
University of Evansville, Indiana, 67
USAir: financial problems, 52, 100–101,
102; history of company, 99–101;
insurance companies, 92, 124, 129, 205;
lawsuits against, 92–93, 123–24,
157–61, 205–7; merger with Piedmont
Airlines (1989), 11, 45, 52, 100, 101;
name change to USAirways (1997), 208;
public relations campaign, 103; responses
to families of crash victims, 50, 109, 111,
125–26; safety campaign, 18, 102; safety
record, 23, 52–53, 83, 101; stock, 101,
143
USAir Flight 405 crash (1993), 56
USAir Flight 427 crash (1994): air traffic
control transcript, 54; bird theories,
66–67, 85, 86–87, 129, 130, 131; "black
boxes," 32, 38–41; bomb theory, 63–64,
108; cockpit tape, 39–40, 48–49,
103–5, 136–37, 146–47, 183–85;
conspiracy theories, 62–64; "fat guy"
theory, 130, 150, 151–52; final NTSB
meeting (March 1999), 200–204; first
anniversary ceremony, 140–43; first NTSB
public hearing (January 1995), 105,
107–12, 125; flight data recorder, 32,
40–42, 83, 88–89; identification of
bodies, 58–61; impact on aviation,
208–9; media coverage, 21–25, 50–53,
63, 186; NTSB probable cause statement,
194, 199, 202; passenger list, 22, 23–24,
26; recovery of wreckage, 32, 43–45,
55–58, 68–69; second NTSB public
hearing (November 1995), 147–50; video
animation, 108, 196; wake turbulence
tests, 146–47; witnesses, 16, 25, 64–65,
128–29
USAir Flight 1016 crash (1994), 18, 50, 58,
92, 101
U.S. Congress: passage of aviation disaster
law, 209; pressure on NTSB, 2
U.S. Department of Commerce: crash
investigations, 33–34, 35

ValuJet crash (1996), 51, 52, 123, 186, 190
Van Bortel, Bonnie, 24, 49, 51
Van Bortel, Brett, 3–4, 20–27, 49–52,
79–83, 180–81, 209–10; final NTSB
meeting on Flight 427, 200, 203–4;
lawsuit, 91–93, 122–25, 157–61, 205–7,
210; memorial services attended, 51–52,
79–80, 140–43; NTSB public hearings,
105, 110; plan for theme restaurant, 158,
181; visits to crash site, 49, 81–82, 106,
142
Van Bortel, Grant, 22
Van Bortel, James, 24
Van Bortel, Joan, 3–4, 6–7, 20–27, 49–52,
79–83, 105–6; identification of remains,
80; insurance policies and workers'
compensation, 93, 124; memorial services
for, 51–52, 79–80, 140–43; scholarship
fund, 125, 207, 210. See also Van Bortel,
Brett, lawsuit
Van Bortel v. USAir Inc., the Boeing Company
and Gerald E. Fox, 93
vertical motion simulator (VMS), 131–32, 132
Vick, Ralph, 170–71
Vogt, Carl, 31–32, 54, 77, 110

Walbert, Cal, 64
Wall Street Journal (newspaper), 180
Washington Post (newspaper), 96, 102, 110,
186
WAVES (computer program), 136, 184
Weaver, Lindsay, 7
Weaver, Scott, 7
Weintraub, Rick, 102
Wheatley, Craig, 24, 49
White House, Washington, D.C., 17
Why Airplanes Crash (Oster, Strong, and
Zorn), 119
WINDFALL (computer program), 55
Winston-Salem, North Carolina: hangar,
130–31, 137; USAir consumer affairs
office, 22
Wolf, Stephen, 208
Wolk, Arthur, 36, 148
Worrell, Dewitt, 126
Wright, Orville, 33
Wright, Wilbur, 33
Wright brothers: plane fatality (1908), 33,
53; use of flight data recorder, 40

Zorn, C. Kurt, 119